REGIONAL POPULATION PROJECTION MODELS

ANDREI ROGERS
University of Colorado, Boulder

SAGE PUBLICATIONS
The Publishers of Professional Social Science
Beverly Hills London New Delhi

For information address:

SAGE Publications, Inc.
275 South Beverly Drive
Beverly Hills, California 90212

SAGE Publications India Pvt. Ltd.
M-32 Market
Greater Kailash I
New Delhi 110 048 India

SAGE Publications Ltd
28 Banner Street
London EC1Y 8QE
England

Printed in the United States of America

Library of Congress Cataloging in Publication Data

Main entry under title:

International Standard Book Number 0-8039-2374-0

Library of Congress Catalog Card No. L.C. 85-050412

FIRST PRINTING

CONTENTS

INTRODUCTION TO THE SCIENTIFIC GEOGRAPHY SERIES

Scientific geography is one of the great traditions of contemporary geography. The scientific approach in geography, as elsewhere, involves the precise definition of variables and theoretical relationships that can be shown to be logically consistent. The theories are judged on the clarity of specification of their hypotheses and on their ability to be verified through statistical empirical analysis.

The study of scientific geography provides as much enjoyment and intellectual stimulation as does any subject in the university curriculum. Furthermore, scientific geography is also concerned with the demonstrated usefulness of the topic toward explanation, prediction, and prescription.

Although the empirical tradition in geography is centuries old, scientific geography could not mature until society came to appreciate the potential of the discipline and until computational methodology became commonplace. Today, there is widespread acceptance of computers, and people have become interested in space exploration, satellite technology, and general technological approaches to problems on our planet. With these prerequisites fulfilled, the infrastructure needed for the development of scientific geography is in place.

Scientific geography has demonstrated its capabilities in providing tools for analyzing and understanding geographic processes in both human and physical realms. It has also proven to be of interest to our sister disciplines, and is becoming increasingly recognized for its value to professionals in business and government.

The Scientific Geography Series will present the contributions of scientific geography in a unique manner. Each topic will be explained in a small book, or module. The introductory books are designed to reduce the barriers of learning; successive books at a more advanced level will follow the introductory modules to prepare the reader for contemporary developments in the field. The Scientific Geography Series begins with several important topics in human geography, followed by studies in other branches of scientific geography. The modules are intended to be used as classroom texts and as reference books for researchers and professionals. Wherever possible, the series will emphasize practical utility and include real-world examples.

We are proud of the contributions of geography, and are proud in particular of the heritage of scientific geography. All branches of geography should have the opportunity to learn from one another; in the past, however, access to the contributions and the literature of scientific geography has been very limited. I believe that those who have contributed significant research to topics in the field are best able to bring its contributions into focus. Thus, I would like to express my appreciation to the authors for their dedication in lending both their time and expertise, knowing that the benefits will by and large accrue not only to themselves but to the discipline as a whole.

—*Grant Ian Thrall*
Series Editor

SERIES EDITOR'S INTRODUCTION

Professor Andrei Rogers, a recognized world leading authority in population futures, turns his attention here to regional population projection. By reading this book you can learn how to do subnational population projections. Follow the presentation with pencil and paper at hand; trace with Andrei Rogers through the mathematical analysis step by step, made clear and relevant by demonstrations of actual projections of population futures at the regional level. The examples show how to calculate numerically regional population growth rates, age compositions, and spatial distributions using data from several developed and less developed countries.

Andrei Rogers demonstrates that projecting population futures at the regional level is both a rewarding intellectual exercise and a powerful technique. Public and private institutions, organizations, and firms require information on potential demographic futures. Public organizations must anticipate future needs and thereby judge whether or not efforts should be launched to alter current population processes and trends. Private firms maximize possible profits by adjusting product lines and shifting distribution networks using information obtained from regional demographic projections.

By disaggregating national populations spatially, Andrei Rogers can analyze the evolution of multiple regional populations, each interconnected by migration flows. The dynamics of the evolution of every subnational human population is governed by the interaction of births, death, and migration. Individuals are born into a population, with the passage of time they age and reproduce, and because of death or outmigration they ultimately leave the population. These events and flows enter into an accounting relationship in which the growth of the regional population is determined by the combined effects of natural increase (births minus deaths) and net migration (inmigrants minus outmigrants).

Andrei Rogers adopts a geographical perspective by considering how fertility, mortality, and migration combine to determine the growth, age composition, and spatial distribution of a national multiregional population; his analysis considers simultaneously: several interdependent subnational population *stocks*; the *events* that alter the levels of such stocks; the aggregate directions *flows* that connect these stocks to form a system of interacting subnational populations.

This module should be of use to those responsible for carrying out regional population projections in public and private organizations such as national, state, and local governments, business firms, foundations, universities, labor unions, social service organizations, and various public interest groups. Students will find that this work by Andrei Rogers contributes a new and significant dimension to human geography and anthropology, sociology and demography, business marketing, regional economics, environmental studies, and city planning.

—*Grant Ian Thrall*
Series Editor

ACKNOWLEDGMENTS

This short monograph could not have been written without the help of a number of individuals and institutions whose assistance over the past several years I wish to gratefully acknowledge. Peer Just, Dimiter Philipov, Friedrich Planck, John Watkins, and Frans Willekens assisted me at different times by writing and executing the various computer programs that produced the numerical results reported here. The International Institute for Applied Systems Analysis in Laxenburg, Austria, and the Institute of Behavioral Science at the University of Colorado, Boulder, provided support and productive environments for these activities. Jeri Futo typed and retyped various versions of the manuscript with good cheer, and Luis Castro, Hazel Morrow-Jones, Chris Rogers, and John Watkins read the entire manuscript and offered useful comments. To all my sincere thanks.

REGIONAL POPULATION PROJECTION MODELS

ANDREI ROGERS
Professor of Geography and
Director, Population Program
Institute of Behavioral Science
University of Colorado, Boulder

1. INTRODUCTION

Population projections are numerical estimates of future demographic totals, usually obtained by the extrapolation of past and current trends. Such calculations are fundamental inputs to social and economic planning. They identify potential demographic futures, anticipate the needs that such futures are likely to create, and provide a basis for judging whether or not efforts should be launched to alter current population processes and trends.

The principal uses of population projections arise in connection with the planning activities of governmental organizations and private firms. National, state, and local governments, business firms, universities, foundations, labor unions, social service organizations, and various public interest groups all use population projections. In addition to various expected population totals, their planning efforts may also require more specialized types of projections, such as expected future numbers of teachers, classrooms, housing units, medical personnel, hospital beds, nursery schools, day care centers, highways, dams—the list goes on and on.

Besides their uses in public and private planning activities, population projections also contribute to the development of a better understanding of demographic phenomena. In particular, they "permit experiments out of which we obtain causal knowledge; they explain data; they focus research by identifying theoretical and practical issues; they systematize comparative study across space and time; they reveal formal analogies between problems that on their surface are quite different; they even help assemble data" (Keyfitz, 1971, p. 573).

Subnational Population Projection: A Geographical Perspective

The subject of this book is subnational population projection. As the modifier subnational suggests, the focus is on the regional populations that collectively make up the national total. National populations are spatially disaggregated, and attention is directed at the evolution of multiple regional populations interconnected by interregional migration flows. It is this orientation that brings forth the particular contribution of geography to demographic analysis.

In the absence of data on births, deaths, and migration, early efforts at population projection necessarily relied on crude methods of extrapolating past observations, usually by fitting simple curves to the data. But a curve may fit observed data for over a century with considerable accuracy and yet fail to predict the situation for the next few years. The generally unsatisfactory results of curve-fitting efforts led to the development of an approach that introduces the behavior of the principal components of population change into the projection exercise—an approach in which trends in fertility, mortality, and migration are taken into account in the projection of population totals. Chapter 1 examines both methods of population projection and also presents the method used by the United Nations to generate urbanization projections.

Chapter 2 considers the consequences for subnational population projection of dividing an aggregate population into spatially distinct, interacting regional populations that exchange migrants in both directions. This permits one to associate gross migration flows with the regional populations that are exposed to the possibility of experiencing them. Gone is the statistical fiction of the net migrant; an outmigrant from one region becomes an inmigrant to another, creating a link between the two regional populations. As a result, instead of considering the dynamics of population redistribution one region at a time, the analyst examines the evolution of a complete system of interacting regional populations simultaneously in a single operation.

Although a number of useful results can be derived without introducing the age dimension into the analysis, serious systematic attention normally is accorded only to demographic projections that disaggregate population totals by age. Such a disaggregation allows one to study the diverse demographic behavior of heterogeneous subpopulations exhibiting differing propensities to experience events and movements. The incorporation of such differences in a formal analysis further illuminates aggregate patterns of demographic behavior. Chapter 3 describes the demographer's classical age-disaggregated single-region approach to population projection. It sets out both the life table and the cohort-survival model and illustrates their application to subnational population projection.

Finally, Chapter 4 integrates the age dimension of the demographer with the locational dimension of the geographer. Populations disaggregated by age and region of residence are advanced over time and across space. Age dynamics are linked with spatial dynamics. Then a reinterpretation of migration between regions as a transition between states of existence generalizes the multiregional projection model into a more general class of models called *multistate* models. The mathematical apparatus for dealing with people moving from one region to another turns out to be formally identical with one dealing with a wide class of transitions that individuals experience during their lifetime: for example, transitions from healthy to ill, from single to married, from employed to unemployed, and from being in school to having graduated. An example of the multiregional/multistate model's generality is illustrated in Chapter 4 below with a population projection that is disaggregated by four different marital statuses and two regions of residence.

Age is usually the most important characteristic of a population in demographic calculations, and to ignore its influence is to invite potentially erroneous findings. Nevertheless, a number of crude but useful results can be derived even if age is disregarded in population projections—as it is in this chapter and the next. Fundamental to these results is the notion of a rate of increase, a subject to which we now turn.

Rates of Increase and Exponential Growth

A population numbering P(t) at a time t and P(t + 1) a year later has exhibited an annual rate of increase of

$$r = \frac{P(t+1) - P(t)}{P(t)} \qquad [1.1]$$

and is said to have been growing at the rate of 100r% per annum. Continued growth at this rate for a decade would mean a total population of

$$P(t + 10) = (1 + r)P(t + 9) = (1 + r)^2 P(t + 8) = \ldots = (1 + r)^{10} P(t) \quad [1.2]$$

or $(1 + r)^{10}$ times the present population.

A convenient means of illuminating the consequences of an unchanging rate of increase is offered by the concept of doubling time. If at the end of ten years a population is $(1 + r)^{10}$ as great as it was a decade before and $(1 + r)^n$ times as large at the end of n years, then the population's doubling time is given by the value of n that satisfies the equation

$$(1 + r)^n = 2 \quad [1.3]$$

Taking natural logarithms and dividing both sides by $\ln(1 + r)$ gives

$$n = \frac{\ln 2}{\ln(1 + r)} \quad [1.4]$$

And recalling that the Taylor series expansion of $\ln(1 + r)$ is $\ln(1 + r) = r - r^2/2 + r^3/3 - \ldots$, we obtain the approximate solution

$$n = \frac{\ln 2}{r} \doteq \frac{0.693}{r} \quad [1.4']$$

when r is small enough for all terms beyond the first to be disregarded in the expansion. Keyfitz and Beekman (1984) show that for commonly observed values of r among human populations (i.e., between 0 and 0.04), a slightly more precise approximation is offered by $n = 0.70/r$, when r is compounded annually. We conclude, therefore, that a population increasing at 2% per annum ($r = 0.02$) doubles in 35 years, at 3% in 23.3 years, and at 4% in 17.5 years.

Interest is said to be compounded annually when a sum invested at the beginning of a year increases to $(1 + r)$ times its value by the end of the year. If the rate of increase r is compounded k times during the year, then the sum at the end will have grown to $[1 + r/k]^k$ times its value at the start of the year. When the rate is compounded continuously, we have

$$\lim_{k \to \infty} \left[1 + \frac{r}{k}\right]^k = e^r \quad [1.5]$$

where e is the base of natural logarithms and is equal to approximately 2.718. Thus a population, P(t), growing at 100r% a year compounded continuously (that is, the k becomes infinitely large), would total

$$P(t + 1) = e^{r} P(t) \quad [1.6]$$

by the end of a year, and over n years would grow to

$$P(t + n) = e^{rn} P(t) \quad [1.6']$$

To reach this same total with annual compounding would require a rate of increase, r* say, that is slightly larger than the corresponding continuous rate r. Specifically, recalling equations 1.2 and 1.6, we observe that

$$(1 + r^*) P(t) = e^{r} P(t) \quad [1.7]$$

whence

$$r^* = e^{r} - 1 \quad [1.8]$$

Thus, for example, the population of Kenya, which has been increasing at about 4.0% per year compounded annually, has been growing at the rate of 3.92% per year compounded continuously. Its doubling time, n, under the latter model is defined by the relation

$$e^{rn} = 2 \quad [1.9]$$

which satisfies equation 1.4′ exactly and yields a value of 17.7 years.

So far we have considered only fixed rates of increase, r. We now turn to time dependent variable rates of increase r(t) and examine the population P(t) that evolves after T years of exposure to such a rate.

Imagine that the time interval of T years is divided into short subintervals dt in length, and suppose that r(t) over the first subinterval is r_0, over the second is r_1, and so on. If P(0) is the initial population total, then $P(0)\, e^{r_0 dt}$ is the total at time dt, and, writing $e^{r_i dt}$ as $\exp[r_i dt]$, gives

$$P(T) = P(0) \exp[r_0 dt] \cdot \exp[r_1 dt] \cdot \exp[r_2 dt] \cdots \exp[r_{T-1} dt] \quad [1.10]$$

$$= P(0) \exp[\Sigma r_i dt]$$

As dt tends to zero, one obtains, in the limit, an integral in place of the summation:

$$P(T) = P(0) \exp \left[\int_0^T r(t)\, dt \right] \qquad [1.11]$$

Taking logarithms of both sides of 1.11 and then differentiating with respect to t yields the following definition of the variable rate of increase:

$$r(t) = \frac{1}{P(t)} \frac{dP(t)}{dt} \qquad [1.12]$$

As Keyfitz (1977) points out, a convenient way of assessing the numerical effect of a variable rate of increase on a population total is to apply the average rate across the entire time interval from zero to T:

$$P(T) = P(0) e^{\bar{r}T} \qquad [1.13]$$

where

$$\bar{r} = \frac{\int_0^T r(t)\,dt}{T} \qquad [1.14]$$

Alternative Projections of National Population Growth and Urbanization

In 1980 the United Nations projected a population of 238 million for Indonesia in the year 2000 (United Nations, 1980). Underlying that projection were the following assumed average annual rates of increase: 2.59% for the period 1975-80, 2.38% for 1980-90, and 1.89% for 1990-2000. This projection adds 16 million more people to the year 2000 total projected three years earlier by the same agency, and it exceeds the corresponding projection carried out by the World Bank in 1979 by some 31 million people. How are we to judge whether the projection is reasonable or not?

The UN estimate of Indonesia's population in 1975 is 136 million. During the preceding five years this population was growing at an average annual rate of 2.6%. Were it to continue to grow at that rate, its total by the year 2000 would be (equation 1.6′):

$$P(2000) = 136\ e^{0.026(25)} = 261 \text{ million}$$

a total that exceeds the 1980 UN projection by some 23 million people.

One can argue that the 261-million figure is undoubtedly too high because the 1970-75 growth rate is likely to decline as national development proceeds. The UN projection assumed a nonlinear pattern of decline to 1.89% by the 1990-2000 decade; alternatively, an assumed linear decline to 2.0% implies instead an average growth rate of 2.3% during the 25-year period and projects a total population of

$$P(2000) = 136\ e^{0.023(25)} = 242 \text{ million}$$

Extrapolating this latter pace of decline into the twenty-first century drops the rate to zero by the year 2083. If one assumes that this rate will remain fixed at zero forever thereafter, then 556 million is the corresponding ultimate zero population growth (ZPG) total. Table 1.1 sets out these totals for Indonesia and compares them with corresponding results for four other Southeast Asian nations.

Moving from national projections to regional ones, we let $P(t)$, $P_u(t)$, and $P_v(t)$ denote, respectively, the total, urban, and rural populations of a country at time t, and let m be the net outmigration rate from rural areas. (We use the letter v as a subscript for rural area variables to avoid having the letter r denote two different attributes: a rate of increase and a rural location.) Assume that the rate of natural increase (birth rate minus death rate) of the urban population is equal to that of the rural population, both of them therefore being equal to the national rate of increase r. (We assume a national population that is undisturbed by international migration.) It follows, then, that

$$P(t) = P(0)e^{rt} \qquad [1.15]$$

Because the growth rate of the rural population will be the rate of natural increase, r, less the rate of net outmigration to urban areas, m:

$$P_v(t) = P_v(0)e^{(r-m)t} \qquad [1.16]$$

and, because the total population is the sum of its rural and urban subpopulations,

$$P_u(t) = P(t) - P_v(t) \qquad [1.17]$$

TABLE 1.1 Historical Population Data and Alternative National Projections (in millions) to the Year 2000 and Beyond

			Alternative Projections						
	Historical Data		Published Projections			Transparent Models*			
Nation	*1950*	*1975*	*UN80*	*UN77*	*World Bank*	*Const. r*	*Decl. r*	*ZPG Yr.*	*ZPG Pop.*
Cambodia	4	8	16	13	16	16	15	2065	28
Indonesia	75	136	238	222	207	261	242	2083	556
Malaysia	6	12	22	20	20	25	22	2056	39
Philippines	21	44	90	83	76	103	87	2037	125
Thailand	20	42	86	76	69	95	81	2039	121

SOURCES: United Nations (1980), World Bank (1979), and Rogers (1981).

*Constant r means the 1970-1975 value given for the country in United Nations (1980); declining r means a linear decline to r = 20 per thousand by the year 2000; continuing this linear decline to zero gives, in the final two columns, the year at which zero population growth (ZPG) first occurs and the ZPG total, respectively.

TABLE 1.2 Average Annual Rural Net Outmigration Rates (per thousand) Implied by Current United Nations Estimates and Projections, 1950-2000

	Historical Data			*UN80 Projection*		
Nation	*1950-60*	*1960-70*	*1970-75*	*1975-80*	*1980-90*	*1990-2000*
Cambodia	0.55	1.12	2.14	2.94	4.52	7.55
Indonesia	2.52	2.95	3.31	4.39	6.42	9.95
Malaysia	6.28	2.38	2.49	4.16	7.09	11.92
Philippines	4.44	3.76	4.32	5.88	8.90	13.55
Thailand	2.30	0.81	0.84	1.83	3.66	7.20

SOURCE: Ledent and Rogers (1979).

Such a simple model of population dynamics was adopted by Keyfitz (1980) to illuminate a number of demographic aspects of the urbanization process. It can also be used to generate projections of urbanization.

Table 1.2 presents rough estimates of the past and future rural net outmigration rates that are implied by the 1980 United Nations urban and rural population projections for the five Southeast Asian nations included in Table 1.1 (Ledent and Rogers, 1979). It shows, for example, that during the 1970-75 period rural areas in Indonesia were losing population at a net annual rate of 3.3 per thousand. Adopting the simplifying assumption that the urban and rural populations were then both exhibiting an annual rate of natural increase that was equal to the national growth rate of 26.0 per thousand, and assuming fixed rates of natural increase and migration over the projection interval, gives (equation 1.16):

$$P_v(2000) = 0.8157\ (136)\ e^{(0.0260-0.0033)25} = 196 \text{ million} \qquad [1.18]$$

where 0.8157 is the fraction of the national population in 1975 that was rural. Earlier we projected the corresponding national total to be

$$P(2000) = 261 \text{ million}$$

thus, using equation 1.17, we have that

$$P_u(2000) = 261 - 196 = 65 \text{ million}$$

a projection that yields an urbanization level of 25.0%.

Relaxing the assumption of fixed rates by allowing r to follow the nonlinear trajectory assumed in the UN projections, while keeping m fixed at 3.3 per thousand, gives Indonesia an urbanization level of 24.9%

TABLE 1.3 Alternative National Urbanization Projections (in percentages) to the Year 2000 and Beyond

			Alternative Projections			
	Historical Data		*Published Projections*		*Transparent Models**	
Nation	*1950*	*1975*	*UN80*	*UN76*	*Constant m*	*Inc. m*
Cambodia	10.21	12.64	23.70	40.00	17.19	34.57
Indonesia	12.41	18.43	32.26	31.44	24.92	39.48
Malaysia	20.37	27.88	41.59	45.08	32.22	46.13
Philippines	27.13	34.30	49.04	50.82	41.03	51.65
Thailand	10.47	13.58	23.18	27.36	15.39	34.61

SOURCES: United Nations (1976, 1980) and Rogers (1981).
*Constant m means the 1970-75 value given for the country in Table 1.2; increasing m means a linear increase to m = 16 by the year 2000. The nonlinear trajectory of r is kept the same as in the United Nations (1980) projections.

by the year 2000 (again using equations 1.16 and 1.17 as above). The 1980 UN projection gives 32.3% for this figure, a consequence of the assumed gradual increase in net rural outmigration to an annual rate of 9.95 per thousand. To bracket this UN projection, we also show in Table 1.3 the corresponding projection with the rural net outmigration rate increasing linearly from its 1970-75 value to 16 per thousand by 1990-2000. This assumption, of course, produces a higher urbanization level than is envisioned in the UN projections—a level of 39.5% to the UN's 32.3%. Analogous findings are obtained for the other four Southeast Asian nations.

The Demographic Sources of Urban Growth

Do cities grow mostly by the surplus of urban births over urban deaths (urban natural increase) or do they grow mostly as a consequence of net inmigration from rural areas? A recent study by the United Nations concluded that urban population growth in the less developed nations results primarily from the natural increase of their urban populations:

> Considering only the most recent observation for a country, an average of 60.7 per cent of growth is attributable to this source, compared with only 39.3 per cent for migration. These figures are nearly reversed for the more developed countries (40.2 and 59.8 percent) [United Nations, 1980, p. 23].

The United Nations' decomposition strives to disentangle the immediate contributions of natural increase and migration to urban population growth by estimating the fraction of today's growth that would be eliminated if rates either of natural increase or of migration were suddenly to drop to zero. But this is a static cross-sectional view, one that ignores the evolution of the changing contributions of migration and of natural increase to urban growth over time. The long-run impacts of current patterns of natural increase and migration on urban population growth and urbanization levels can be conveniently assessed by population projection.

Without a city population there obviously cannot be any urban natural increase; and for some time after the establishment of a city, when its population is still relatively small in size, the contribution of urban net inmigration is likely to exceed that of urban natural increase. At the other extreme, when a nation is mostly urbanized, the outmigration of its rural population can contribute little to urban growth. Between these two extremes comes a time at which the contribution of natural increase begins to dominate that of net inmigration.

Imagine a hypothetical population, initially entirely rural, that experiences an annual rate of natural increase of r and a net rural outmigration rate of m. Recalling equations 1.12 and 1.15-1.17, one may establish that the rate of growth of the urban population is the sum of the contribution of urban natural increase and the contribution of urban net inmigration (Keyfitz, 1980):

$$r_u = \frac{1}{P_u(t)} \frac{d\,P_u(t)}{dt} = \frac{d}{dt} \ln P_u(t) = r + \frac{m}{e^{mt} - 1} \qquad [1.19]$$

The first term, r, is the assumed rate of natural increase of the urban population; thus the second term, $m/(e^{mt} - 1)$, must be its rate of increase through migration. Dividing the second by the first gives

$$R(t) = \frac{m}{r(e^{mt} - 1)} \qquad [1.20]$$

the ratio of the contribution of migration to natural increase.

At what point does the contribution of natural increase to urban growth first begin to exceed that of urban net inmigration? The former overtakes the latter immediately after R(t) reaches unity, an event that takes place when

$$\frac{m}{r} = e^{mt} - 1 \qquad [1.21]$$

$$\ln\left[\frac{m}{r} + 1\right] = mt \qquad [1.22]$$

or

$$t = \frac{1}{m} \ln\left[\frac{m}{r} + 1\right] \qquad [1.23]$$

Keyfitz (1980) calls t *the crossover point* and observes that the faster the rate of population growth, the sooner the crossover, and the larger the rural net outmigration, the sooner will natural increase exceed migration as the principal contributor to urban population growth.

Denoting the crossover moment by t_c, we note that at this point the ratio of the urban to rural population is

$$S(t_c) = \frac{P_u(t_c)}{P_v(t_c)} = e^{mt_c} - 1$$

$$= \exp\left\{m\left[\frac{1}{m} \ln\left[\frac{m}{r} + 1\right]\right]\right\} - 1$$

$$= \frac{m}{r} \qquad [1.24]$$

the fraction of the national population that is urban is

$$U(t_c) = \frac{P_u(t_c)}{P_u(t_c) + P_v(t_c)} = \frac{S(t_c)}{1 + S(t_c)}$$

$$= \frac{m}{r + m} \qquad [1.25]$$

and, recalling 1.19, the growth rate of the urban population is

$$r_u = r + \frac{m}{e^{mt_c} - 1} = r + \frac{m}{\exp\{m(1/m)\ln[(m/r) + 1]\} - 1} = 2r \qquad [1.26]$$

The above results can be derived directly. At the crossover point, the contribution of urban net inmigration, $mP_v(t_c)$, is equal to that of natural increase, $rP_u(t_c)$. Therefore,

$$S(t_c) = \frac{P_u(t_c)}{P_v(t_c)} = \frac{m}{r} \qquad [1.27]$$

At this moment the urban rate of natural increase is r and that of urban net inmigration is equal to it; hence the growth rate of the urban population is 2r.

Imagine again a hypothetical population, initially entirely rural, for which the rate of natural increase is fixed at 3% per year and the annual rate of net outmigration from rural areas is set at 2%. The ratio of net inmigration to natural increase in the urban population will be 3.01, 1.36, 0.81, and 0.39, after 10, 20, 30, and 50 years, respectively. The crossover point is reached

$$t_c = \frac{1}{0.02} \ln(1 + 0.02/0.03)$$

$$= 25.5 \text{ years}$$

after the start of the urbanization process, when the urban population is 40% of the total and growing at a rate of 6% a year.

The urban population of India in 1970 was increasing by about 3.7% a year (Rogers 1978b; Appendix A). This urban growth rate was the outcome of a 2% rate of natural increase and a net rural outmigration rate of 0.5%. Thus, substituting into equation 1.23, gives

$$t_c = \frac{1}{0.005} \ln\left[\frac{0.005}{0.020} + 1\right]$$

$$= 44.6 \text{ years}$$

India's urban population in 1970 was just about to start growing more from natural increase than from net inmigration; the crossover point was reached that year. To see this, we note that India's observed urbanization level in 1970 was U = 0.20, which is precisely the level reached at the moment of crossover (equation 1.25):

$$U(t_c) = \frac{0.005}{0.020 + 0.005} = 0.20$$

The natural increase of rural populations exceeds that of urban populations in most parts of the world. Thus a greater degree of realism may be obtained by distinguishing these two rates in the above model. Such an extension appears in Keyfitz (1980).

Logistic Growth

We have seen, in Section 1.2, that if

$$\frac{d}{dt} P(t) = r\, P(t) \qquad [1.28]$$

then

$$P(t + 1) = e^r\, P(t) \qquad [1.29]$$

that is, a constant rate of population growth gives rise to an exponentially increasing population. Introducing a "ceiling" (a maximum value to ultimate population size, say K) by adding to 1.28 a dampening factor $[1 - P(t)/K]$ that gradually depresses the exponential growth component as the ceiling is approached, gives

$$\frac{d}{dt} P(t) = \frac{r}{K} P(t)\, [K - P(t)] \qquad [1.30]$$

Equation 1.30 states that population increase is proportional to the total already attained, P(t), and to the remaining "distance" between that total and the assumed population ceiling, K. Note that the right-hand side of 1.30 is zero when P(t) = K; the population enters a zero growth regime at that point.

Reexpressing 1.30 as (Keyfitz, 1977):

$$\left[\frac{1}{P(t)} + \frac{1}{K - P(t)}\right] d\, P(t) = r dt \qquad [1.31]$$

and integrating, yields

$$\ln\left[\frac{P(t)}{K - P(t)}\right] = rt + c \qquad [1.32]$$

Taking exponentials and then solving for P(t) gives

$$P(t) = e^{rt+c}\ [K - P(t)] \qquad [1.33]$$

whence

$$P(t) = \frac{Ke^{rt+c}}{1 + e^{rt+c}} \qquad [1.34]$$

or, dividing both the numerator and the denominator by e^{rt+c},

$$P(t) = \frac{K}{1 + e^{-rt-c}} \qquad [1.35]$$

Equation 1.35 defines the curve known as the logistic growth function. It appears in many guises in the literature; two of its most common alternative forms are

$$P(t) = \frac{K}{1 + Be^{-rt}} \qquad [1.36]$$

where $B = e^{-c}$, or

$$P(t) = \frac{1}{A + Ce^{-rt}} \qquad [1.37]$$

when the ceiling is assumed to equal unity by setting A = 1/K and C = B/K.

Figure 1.1 illustrates the fit of the logistic curve to a scatter of observations exhibiting the association of national levels of urbanization with per capita gross national product (GNP) for a number of countries. Note that the logistic curve of urbanization exhibits a 50%

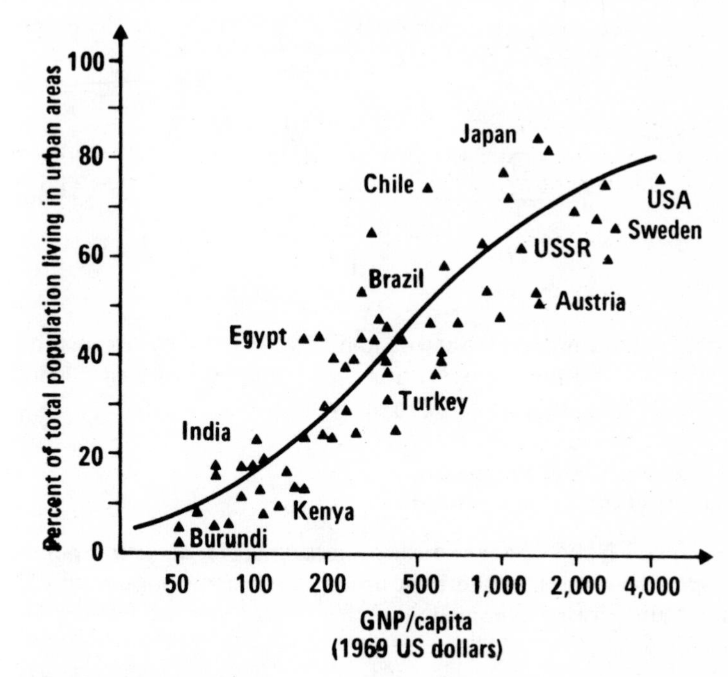

SOURCE: Ledent (1980).

Figure 1.1 The Logistic Association Between the Degree of Urbanization and Per Capita GNP for World Bank Member-Countries in 1970

urban national population at a per capita GNP of \$500 measured in 1969 U.S. dollars.

Crude estimates of the parameters of the logistic curve may be obtained by simple linear regression, inasmuch as equation 1.36 may be expressed as

$$\frac{1}{P(t)} = \frac{1 + Be^{-rt}}{K} = \frac{1}{K} + \frac{B}{K} e^{-rt} \qquad [1.38]$$

whence

$$\ln\left[\frac{K}{P(t)} - 1\right] = \ln B - rt \qquad [1.39]$$

Thus the input data for regression are a set of paired values for t and $\ln[K/P(t) - 1]$.

If estimates for $P(t_1)$, $P(t_2)$, and $P(t_3)$ are available at equidistant times t_1, t_2, and $t_3 = (2t_2 - t_1)$, with $P(t_1) < P(t_2) < P(t_3)$, then the population ceiling K is given by (Keyfitz, 1977):

$$K = \frac{\frac{1}{P(t_1)} + \frac{1}{P(t_3)} - \frac{2}{P(t_2)}}{\frac{1}{P(t_1)P(t_3)} - \frac{1}{[P(t_2)]^2}} \qquad [1.40]$$

provided both the numerator and denominator in the above expression for K are positive. In some applications, for example, projected urbanization paths, K is often set equal to unity.

The Urban and Rural Population Projections of the United Nations

The United Nations' method for projecting urban and rural population growth assumes that both the urban and the rural populations of a country are growing exponentially:

$$P_u(t + n) = P_u(t)e^{r_u n} \qquad [1.41]$$

$$P_v(t + n) = P_v(t)e^{r_v n} \qquad [1.42]$$

The ratio of these two populations, denoted by S(t), must then also be growing exponentially:

$$S(t + n) = \frac{P_u(t + n)}{P_v(t + n)} = \frac{P_u(t)e^{r_u n}}{P_v(t)e^{r_v n}} = S(t)e^{(r_u - r_v)n} = S(t)e^{gn} \qquad [1.43]$$

where $g = r_u - r_v$ is the urban-rural growth rate difference (URGD in United Nations parlance).

Recalling that

$$U(t) = \frac{P_u(t)}{P_u(t) + P_v(t)} = \frac{S(t)}{1 + S(t)} \qquad [1.44]$$

we establish through substitution that

$$U(t+n) = \frac{1}{1 + [1/S(t)]\,e^{-gn}} \qquad [1.45]$$

a form of the logistic function that appears in equation 1.36, with r = g, t = n, K = 1, and B = 1 / S(t). Thus, if the urban-to-rural population ratio is growing exponentially, the urban proportion of the national population is increasing according to the logistic growth curve.

If the most recently observed value of g were assumed to remain constant, then urban-rural population projections could be carried out simply by applying equation 1.45 to independently projected national population totals. However, a constant urban-rural growth rate difference generally produces values of U(t) that are too high (United Nations, 1980), especially in those countries with high observed values of g and initial proportions urban.

The overwhelmingly important source of differences between urban and rural population growth rates is net rural outmigration. As a country urbanizes and empties out its village populations, the pool of potential migrants to urban areas declines as a fraction of the urban population at the same time that the pool of potential migrants to rural areas increases as a fraction of the rural population. Consequently, it is reasonable to expect g to decline as U(t) rises.

Recognizing this evolution, the most recent urban and rural projections issued by the United Nations reflect a dampening mechanism that ensures a decline in urban-rural growth rate differences as levels of urbanization increase. Specifically, drawing on cross-sectional data for 110 countries with over 2 million in population, collected in the 1960 and 1970 rounds of censuses, United Nations demographers estimated the following simple linear relationship (United Nations, 1980 and 1984):

$$g_h = 0.044 - 0.028\,U(t_0) \qquad [1.46]$$

where g_h is the "hypothetical" value of the urban-rural growth rate difference and $U(t_0)$ is the initial proportion urban. According to this equation, when a country is 10% urban, the expected value of g_h is 0.041; when it is 90% urban, g_h is only 0.019. Inasmuch as few countries have actually experienced these expected values, the United Nations' method allows local conditions to be reflected early in the projection period by incorporating a weighting procedure in which g is gradually forced to converge to g_h over time.

TABLE 1.4 Percentage of Population Living in Urban Areas of Major Areas and Regions, 1950-2000

	1950	*1960*	*1970*	*1975*	*1980*	*1990*	*2000*
World total	28.95	33.89	37.51	39.34	41.31	45.88	51.29
More developed regions	52.54	58.73	64.68	67.49	70.15	74.87	78.75
Less developed regions	16.71	21.85	25.82	28.03	30.53	36.46	43.46
Africa	14.54	18.15	22.85	25.67	28.85	35.70	42.49
Latin America	41.18	49.45	57.37	61.21	64.74	70.70	75.21
Northern America	63.84	67.09	70.45	71.99	73.66	77.20	80.76
East Asia	16.72	24.71	28.61	30.70	33.05	38.63	45.43
South Asia	15.65	17.80	20.45	22.02	23.95	29.10	36.13
Europe	53.70	58.42	63.94	66.45	68.83	73.25	77.11
Oceania	61.24	66.22	70.77	73.35	75.93	80.37	82.97
USSR	39.30	48.80	56.70	60.90	64.77	71.28	76.06

SOURCE: United Nations (1980).

In the United Nations' method the projected value of g, denoted by g(t), is defined as the linear weighted combination of the last observed value g_0 and the country's "hypothetical" value, g_h:

$$g(t) = W_1(t)g_0 + W_2(t)g_h \qquad [1.47]$$

with the weights satisfying the constraint $W_1(t) + W_2(t) = 1$.

The values assigned to the weights over time are

Time Interval	$W_1(t)$	$W_2(t)$
1975-1980	0.8	0.2
1980-1985	0.6	0.4
1985-1990	0.4	0.6
1990-1995	0.2	0.8
1995-2000	0	1
⋮	⋮	⋮
2020-2025	0	1

Thus during the early parts of the projection more weight is given to the last observed value of g; with the passage of time, the country's "hypothetical" value g_h is given a heavier weighting. Ultimately the projected g converges to the hypothetical g_h, but the crude and arbitrary nature of this weighting procedure can introduce sudden unexplained shifts in the projected growth rate.

The national urbanization projections set out earlier in Table 1.3 were produced by this United Nations method. Table 1.4 presents the associated global and regional projections, which were obtained by aggregating the appropriate country totals.

This chapter has introduced a number of fundamental concepts in subnational population projection: rate of increase, exponential growth, logistic growth, and the ratio of the contributions of migration and of natural increase to subnational population growth. A more advanced exposition of these same concepts can be found, for example, in the works of such mathematical demographers as Keyfitz (1977, 1980). We now shall consider some of these same concepts in the context of a spatial system of several interacting subpopulations.

2. SPATIAL POPULATION DYNAMICS: LOCATION WITHOUT AGE

Changes in the size and composition of a human population residing in a territorial unit with fixed boundaries are determined by the interaction of births and deaths to residents and the migration across those boundaries of both residents and nonresidents. When boundaries change over time, as they do with populations classified as urban and rural, territorial reclassification also becomes a contributor to change. Here we shall follow the standard practice of international agencies, such as the United Nations, and include the effects of reclassification together with those of internal migration, distinguishing them from natural increase, which will also be assumed (as in standard UN practice) to include the effects of international migration.

To introduce the spatial dimension into the analysis, we disaggregate the national population into regional subpopulation and replace net migration rates with origin-destination-specific gross migration rates. The net rates express differences between arrivals and departures as a fraction of the single population experiencing both; the gross rates, on the other hand, express the total number departing from one subpopulation and arriving at another as a fraction of the origin subpopulation. This revised perspective brings with it a fundamental reorientation in subnational population analysis. It introduces a view of all regional populations as mutually interacting components of a multiregional spatial system. Instead of addressing the evolution of each regional population one at a time, the multiregional perspective examines the evolution of all regional populations at the same time. This reformulation leads to several advantages in the quantitative study of migration patterns and the ways in which they affect spatial population dynamics.

Net migration rates are imperfect indices of movement propensities because they also reflect the sizes of the populations from which they derive. For example, if the gross rates of migration between urban and rural areas of a nation are held constant, the corresponding urban net migration rate will change over time with shifts in the relative population totals in the two areas. Accordingly, one's inferences about changes in net migration patterns over time will confound the impacts of migration propensities with those of changing population stocks.

Subnational population projection models that adopt net migration rates cannot keep track of population subgroups that are distinguished by places of birth or by previous regions of residence. Such disaggregations are possible only with models that use gross migration rates. Additional disaggregations to reduce other forms of heterogeneity among migrants may further illuminate the evolution of particular spatial population distributions. For example, if the experience of an event, such as migration, produces a "learning" effect that raises the chances for experiencing another, or if the continued nonexperience of the event produces a form of "cumulative inertia" that lowers the probability of migration, then a disaggregation that reflects some of these effects may be desirable. Only a model based on gross migration flows is capable of incorporating such disaggregations.

In short, a focus on gross instead of net migration flows more clearly identifies the patterns, illuminates the dynamics, and enhances the understanding of demographic processes that occur in multiple interacting populations. Distinguishing between flows and changes in stocks reveals patterns that otherwise may be obscured; focusing on flows into and out of a regional population exposes dynamics that otherwise may be hidden; and linking explanatory hypotheses to appropriately disaggregated gross flows permits a more accurately specified projection model. To introduce the multiregional perspective we begin with the simplest possible model—one with only two subnational aggregate populations: urban and rural.

Urbanization in the Soviet Union

The urban population of the Soviet Union was increasing by about 2.5% a year in 1970 (Rogers, 1978a). The urban growth rate, r_u, was the outcome of a birth rate, b_u, of 17 per 1000; a death rate, d_u, of 8 per 1000; an inmigration rate, i_u, of 27 per 1000; and an outmigration rate, o_u, of 11 per 1000. Expressing these rates on a per capita basis leads to the fundamental identity

$$\begin{aligned} r_u &= b_u - d_u + i_u - o_u \\ &= 0.017 - 0.008 + 0.027 - 0.011 \\ &= 0.025 \end{aligned} \qquad [2.1]$$

The corresponding identity for the rural population was

$$r_v = b_v - d_v + i_v - o_v \quad [2.2]$$

$$= 0.019 - 0.009 + 0.014 - 0.035$$

$$= -0.011$$

The total national population of the USSR in 1970 was about 242 million, of which roughly 136 million (56%) was classified as urban. Multiplying this latter total by the urban growth rate gives

$$136(0.025) = 3.40 \text{ million}$$

as the projected increase for 1971. An analogous calculation for the rural population yields

$$106(-0.011) = -1.17 \text{ million}$$

for the corresponding projected decrease in the rural population. These changes imply, for 1971, an urban population of 139 million, a rural population of 105 million, and a rate of national population increase of

$$\frac{136(0.025) + 106(-0.011)}{242} = 0.56(0.025) + 0.44(-0.011) = 0.009$$

that is,

$$r = 0.56\, r_u + 0.44\, r_v = 0.009$$

where 0.56 is the fraction urban and 0.44 is the fraction rural.

A uniregional perspective of urban growth in the Soviet Union would describe the dynamics of urbanization by focusing on the natural increase and net migration components of the 0.025 rate of urban growth; that is, 0.009 and 0.016, respectively. On the assumption that these rates remain fixed, a ten-year projection gives a 1980 urban population of

$$P_u(1980) = (1.025)^{10}\, 136 = 174.09 \text{ million}$$

Subtracting this quantity from the corresponding projected national total of

$$P(1980) = (1.009)^{10}\, 242 = 264.68 \text{ million}$$

gives, as a residual, a projected rural population of

$$P_v(1980) = 264.68 - 174.09 = 90.59 \text{ million}$$

Reversing the order and projecting the rural population first gives

$$P_v(1980) = (0.989)^{10}\ 106 = 94.90 \text{ million}$$

and, as a residual, a projected urban population of

$$P_u(1980) = 264.68 - 94.90 = 169.78 \text{ million}$$

Finally, an alternative uniregional formulation is one that projects the national total as the sum of the urban and rural projections. In this perspective, the projected national total is

$$174.09 + 94.90 = 268.99 \text{ million}$$

A uniregional projection that relies on the notion of a net migration rate can ultimately produce a fundamental inconsistency. For example, fixed rates of rural outmigration eventually will empty out the rural areas, but net inmigration to the urban areas will continue unabated in the uniregional model. A logical escape from this built-in inconsistency of the uniregional perspective is the adoption of a *biregional* model in which rates of migration reflect the populations that are exposed to the possibility of migrating. The urban rate of natural increase is the same as before, but the urban net migration rate m_u of 0.016 is now decomposed into its in- and outmigration components:

$$m_u = i_u - o_u = 0.027 - 0.011 = 0.016 \qquad [2.3]$$

The flow of inmigrants into urban areas is equal to o_vP_v, the flow out of rural areas. Dividing this quantity by the urban population gives the urban inmigration rate i_u, which is calculable as the product of a rural outmigration rate o_v of 0.035 and the ratio of rural to urban population P_v/P_u of 0.78. Thus

$$m_u = o_v\left(\frac{P_v}{P_u}\right) - o_u = 0.035(0.78) - 0.011 = 0.016 \qquad [2.4]$$

and urban population growth may be described by an accounting equation that involves both the receiving and sending populations:

$$P_u(t + 1) = (1 + b_u - d_u - o_u)P_u(t) + o_vP_v(t) \qquad [2.5]$$

Equation 2.5 states that next year's projected urban population total is calculated by adding to this year's urban population (1) the increment due to urban natural increase, (2) the decrement due to the outmigration to rural areas, and (3) the increment due to the inmigration from rural areas. Substituting in the rates for the Soviet Union gives the accounting identity

$$\begin{aligned} P_u(1971) &= (1 + 0.017 - 0.008 - 0.011) P_u(1970) + 0.035\ P_v(1970) \\ &= 0.998(136) + 0.035(106) \\ &= 139.44 \text{ million} \end{aligned} \quad [2.6]$$

An analogous equation for the rural population yields

$$\begin{aligned} P_v(1971) &= 0.011\ P_u(1970) + (1 + 0.019 - 0.009 - 0.035) P_v(1970) \\ &= 0.011(136) + 0.975(106) \\ &= 104.85 \text{ million} \end{aligned} \quad [2.7]$$

Assuming, once again, that the various rates remain unchanged over a ten-year period gives

$$P_u(1980) = 168.53 \text{ million}$$

$$P_v(1980) = \ \ 97.21 \text{ million}$$

and

$$P(1980) = 168.53 + 97.21 = 265.74 \text{ million}$$

Apart from rounding errors, the biregional model generates the same 1971 projection as do the uniregional models; however, its ten-year projection differs from all three uniregional projections. Moreover, its long-run projection does not locate the entire national population into urban areas. Its stable distribution accords rural areas about a fourth (0.243) of the total stable national population.

The differences between the uniregional and biregional projections are due to bias, and the principal cause of this bias may be shown to be a consequence of treating migration as a net flow. To see this more clearly, consider how the migration specification is altered when the biregional model is transformed into a uniregional model. The accounting relationship in equation 2.5, which involved both the urban and the rural populations, now is replaced by an expression that involves only the former:

$$P_u(t+1) = (1 + b_u - d_u - o_u)P_u(t) + \left[o_v \frac{P_v(t)}{P_u(t)}\right]P_u(t) \qquad [2.8]$$

$$= (1 + b_u - d_u - o_u + i_u)P_u(t)$$

$$= (1 + b_u - d_u + m_u)P_u(t) = (1 + r_u)P_u(t)$$

where

$$m_u = i_u - o_u = \text{urban net inmigration rate} \qquad [2.9]$$

$$i_u = o_v \frac{P_v(t)}{P_u(t)} = o_v \left[\frac{1 - U(t)}{U(t)}\right] \qquad [2.10]$$

and U(t) is the fraction of the total national population that is urban at time t. In the ten-year projection, all annual rates are assumed to be fixed. But i_u, and therefore also m_u, depend on U(t), which varies during the ten years and creates a bias.

Bias and inconsistency may result from viewing biregional (and, by extension, multiregional) population systems through a uniregional perspective. Expressing migration's contribution to regional population growth solely in terms of the population in the region of destination can lead to over- or underprojection and introduce inconsistencies in long-run projections. These problems of bias and inconsistency become even more important when age composition is taken into account, as we shall see in Chapter 4.

Disaggregated Projections

The discussion until now has revolved around uniregional versus biregional models as alternative means for accomplishing the same end: a projection of a nation's urban and rural populations. We now turn to a consideration of ends that can be achieved only with the use of biregional and multiregional models: projections disaggregated by residence-duration status, for example. Because uniregional models do not focus on gross migration flows, they cannot follow the migration paths of particular population groups over time. Biregional and multi-regional models, on the other hand, can identify the life histories of such groups, and this gives them a decisive advantage over uniregional models.

Projections disaggregated by residence-duration status have both a retrospective and a prospective aspect. For example, recalling our earlier projections of the urban and rural populations of the Soviet Union, we may wish to identify how many of the projected urban residents were living in rural areas at the start of the projection period (i.e., in 1970). Or we may be interested in determining what fraction of the projected urban dwellers were born in rural areas (i.e., are "alien" residents) and what proportion are urban "natives."

Prospectively, we may ask, for example, what proportion of the 1970 Soviet rural population will be living in urban areas in the year 2000. To answer this question, we begin by dividing the resident urban population into natives and aliens:

$$\text{residents} = \text{natives} + \text{aliens}$$

$$P_u(t) = {}_uP_u(t) + {}_vP_u(t) \qquad [2.11]$$

where the additional subscript on the left of the population variable denotes the region of birth (the right subscript denotes the region of residence, as before).

The accounting relationship for projecting urban populations was given earlier as equation 2.5. The same equation may be used for projecting the urban native population simply by introducing the place of birth subscript and adding the new births of urban residents during the year to the native population:

$${}_uP_u(t + 1) = (1 + b_u - d_u - o_u)\,{}_uP_u(t) + b_u\,{}_vP_u(t) + o_v\,{}_uP_v(t) \qquad [2.12]$$

Analogous relationships may be defined for ${}_vP_u(t)$, ${}_uP_v(t)$, and ${}_vP_v(t)$. It is assumed that natives and aliens experience the same fertility, mortality, and migration rates—namely, those prevailing at their region of residence. Because all births to alien migrants are added to the native population stock, the equations for ${}_vP_u(t)$ and ${}_uP_v(t)$ contain no birth rates. For example,

$${}_vP_u(t + 1) = (1 - d_u - o_u)\,{}_vP_u(t) + o_v\,{}_vP_v(t) \qquad [2.13]$$

For illustrative purposes, assume that a half of the Soviet Union's 1970 urban population was born in rural areas and that one-tenth of its rural population was born in urban areas. Then

$${}_uP_u(1971) = 0.998(68) + 0.017(68) + 0.035(10.6) = 69.4 \text{ million}$$

$${}_vP_u(1971) = 0.981(68) + 0.035(95.4) = 70.0 \text{ million}$$

$${}_uP_v(1971) = 0.011(68) + 0.956(10.6) = 10.9 \text{ million}$$

$${}_vP_v(1971) = 0.011(68) + 0.019(10.6) + 0.975(95.4) = 94.0 \text{ million}$$

Recalling the "row-times-column" rule of matrix multiplication, we re-express the above four equations in matrix form:

$$\begin{bmatrix} {}_uP_u(1971) \\ {}_vP_u(1971) \\ {}_uP_v(1971) \\ {}_vP_v(1971) \end{bmatrix} = \begin{bmatrix} 0.998 & 0.017 & 0.035 & 0 \\ 0 & 0.981 & 0 & 0.035 \\ 0.011 & 0 & 0.956 & 0 \\ 0 & 0.011 & 0.019 & 0.975 \end{bmatrix} \begin{bmatrix} 68 \\ 68 \\ 10.6 \\ 95.4 \end{bmatrix}$$

Continuing this projection forward for another 29 years yields

$${}_uP_u(2000) = 123.6 \qquad {}_ur_u(2000) = 0.018$$

$${}_vP_u(2000) = 103.0 \qquad {}_vr_u(2000) = 0.006$$

$${}_uP_v(2000) = 21.1 \qquad {}_ur_v(2000) = 0.021$$

$${}_vP_v(2000) = 72.3 \qquad {}_vr_v(2000) = -0.004$$

$${}_uU(2000) = 0.39 \qquad {}_vU(2000) = 0.32$$

where

$$U(2000) = {}_uU(2000) + {}_vU(2000)$$

and

$${}_uU(2000) = \frac{123.6}{123.6 + 103.0 + 21.1 + 72.3} = 0.39$$

The same result may be obtained in a single matrix multiplication if the matrix of growth rates is first raised to the thirtieth power, as is

$$\begin{bmatrix} 123.63 \\ 102.95 \\ 21.06 \\ 72.31 \end{bmatrix} = \begin{bmatrix} 1.0570 & 0.4145 & 0.6256 & 0.1775 \\ 0.0168 & 0.6593 & 0.1342 & 0.5823 \\ 0.1885 & 0.0476 & 0.3374 & 0.0150 \\ 0.0503 & 0.1912 & 0.2370 & 0.5594 \end{bmatrix} \begin{bmatrix} 68 \\ 68 \\ 10.6 \\ 95.4 \end{bmatrix}$$

2.1A. Projection of the urban and rural populations to the year 2000, disaggregated by place of birth.

$$\begin{bmatrix} 226.58 \\ 93.37 \end{bmatrix} = \begin{bmatrix} 1.0738 & 0.7598 \\ 0.2388 & 0.5745 \end{bmatrix} \begin{bmatrix} 136 \\ 106 \end{bmatrix}$$

2.1B. Projection of the urban and rural populations to the year 2000.

$$\begin{bmatrix} 47.50 \\ 34.21 \end{bmatrix} = \begin{bmatrix} 0.6425 & 0.4481 \\ 0.1408 & 0.3227 \end{bmatrix} \begin{bmatrix} 0 \\ 106 \end{bmatrix}$$

2.1C. Survivors of the 1970 rural population in the year 2000.

Figure 2.1 Alternative Projections of the 1970 Urban and Rural Populations of the Soviet Union to the Year 2000

demonstrated in Figure 2.1A. The growth matrix of that multiplication may be aggregated to produce urban-rural projections without reference to place of birth (Figure 2.1B). An analogous matrix, with the contribution of fertility deleted (Figure 2.1C), gives, for example, the proportion of the initial rural population that one may expect to find in urban areas in the year 2000 (i.e., 47.50/106 = 0.45).

A number of interesting conclusions may be drawn at this point. First, on 1970 rates, roughly one-fifth of the urban population in the year 2000 will consist of people who lived in rural areas in 1970; and, given our earlier assumptions regarding place of birth, 45% of the urban population will be made up of rural born. Second, although the national population will be growing at an annual rate of just under 1%, the growth rate of aliens in the rural areas and of natives in the urban regions will be about twice as high. Finally, as is to be expected in a national population experiencing high levels of rural to urban migration, the only declining rate of growth is exhibited by the rural native population.

The Sources of Urban Growth Revisited

Imagine a hypothetical population, initially entirely rural, that is subjected to the regime of growth exhibited by India in 1970 (Appendix A). Table 2.1A.1 shows that after 30 years the urban population is 15.5% of the national total and growing at 4.7% per annum. The rate of urban net inmigration at that moment is 2.7%, and its contribution as a source of urban growth is (2.7/4.7)100 = 58.1%. Twenty years later the urban fraction increases to 22.0% and migration's contribution falls to 41.9%. The crossover point is passed after 39 years, when the urban fraction is 18.7%. Note that this hypothetical population, starting its evolution as an entirely rural population, ultimately stabilizes. This is simply a consequence of what is known in demography as *strong ergodicity*, the tendency of an observed population to "forget its past" eventually as it is projected for a long period into the future under fixed rates of natural increase and migration. Such "horizon-year" projections allow one to contrast two regimes of growth without confounding their impacts with different starting conditions, such as India's initial 20% urban to the Soviet Union's 56% in 1970.

Table 2.1 suggests that India's urban population in 1970 was growing more due to natural increase than to net migration because it passed its crossover point when it was 18.7% urban some time ago. (Compare this result with the corresponding uniregional finding set out at the end of Chapter 1. Why the difference?) The urban population in the Soviet Union in 1970, on the other hand, was growing more as a result of net migration than of natural increase because it still was about 9-10 years short (on 1970 rates) of reaching the 63.2% urban level associated with its crossover point.

The crossover point for the population with the Soviet Union's growth regime occurs at about the same time as with India's—that is, after 39.5 years—but it is experienced by a national population that is much more urban. Tables 2.1A.2 and 2.1B.2 show why. Lowering rates of natural increase delays the crossover point, but raising net rates of urban inmigration hastens its occurrence. Combining India's natural increase with the Soviet Union's higher rates of rural to urban migration reduces the time to the crossover from 39 to 27 years. Replacing these migration rates with India's in the Soviet Union's growth regime delays the crossover by over 20 years.

Table 2.1 indicates that the principal effect of migration is to determine the level of urbanization, whereas that of natural increase is to establish the urban growth rate. Despite differences in migration rates, India's natural increase rate ultimately produces an urban population

TABLE 2.1 Aggregated Projections of Hypothetical Populations Initially Entirely Rural and Exposed to Different Regimes of Growth

A. Hypothetical Population: India

1. India's Growth Regime					2. India's Natural Increase Rates with Soviet Union's Migration Rates			
$b_u - d_u = 20x10^{-3}$	$b_v - d_v = 22x10^{-3}$	$o_u = 10x10^{-3}$	$o_v = 7x10^{-3}$		$b_u - d_u = 20x10^{-3}$	$b_v - d_v = 22x10^{-3}$	$o_u = 11x10^{-3}$	$o_v = 35x10^{-3}$
$100U$	r_u	m_u	$m_u/r_u \times 100$	T	$100U$	r_u	m_u	$m_u/r_u \times 100$
0	0	0	0	0	0	0	0	0
3.2	0.215	0.195	90.9	5	15.5	0.198	0.178	90.1
6.1	0.114	0.094	82.8	10	27.7	0.099	0.079	80.3
15.5	0.047	0.027	58.1	30	55.9	0.036	0.016	45.3
22.0	0.034	0.014	41.9	50	67.1	0.025	0.006	22.8
37.7	0.021	0.001	6.0	Stability	74.7	0.020	0.001	2.5

B. Hypothetical Population: Soviet Union

1. Soviet Union's Growth Regime					2. Soviet Union's Natural Increase Rates with India's Migration Rates			
$b_u - d_u = 9x10^{-3}$	$b_v - d_v = 10x10^{-3}$	$o_u = 11x10^{-3}$	$o_v = 35x10^{-3}$		$b_u - d_u = 9x10^{-3}$	$b_v - d_v = 10x10^{-3}$	$o_u = 10x10^{-3}$	$o_v = 7x10^{-3}$
$100U$	r_u	m_u	$m_u/r_u \times 100$	T	$100U$	r_u	m_u	$m_u/r_u \times 100$
0	0	0	0	0	0	0	0	0
15.7	0.184	0.175	95.1	5	3.3	0.201	0.192	95.6
28.1	0.087	0.078	89.7	10	6.2	0.101	0.092	91.2
56.7	0.024	0.015	63.1	30	15.9	0.035	0.026	74.4
67.9	0.014	0.005	36.6	50	22.7	0.022	0.013	59.5
75.3	0.009	0.000	1.7	Stability	39.6	0.009	0.000	4.1

SOURCE: Rogers, A. 1982a. Sources of Urban Population Growth and Urbanization, 1950-2000: A Demographic Accounting. *Economic Development and Cultural Change 30* (3): 483-506 (Chicago: University of Chicago Press).

growing at 2% per year; the Soviet Union's gives rise to urban growth at roughly half that rate. Despite differences in rates of natural increase, the Soviet Union's migration rates generate a national population that ultimately is three-fourths urban, whereas those of India produce an urban fraction that ultimately is just under 40%.

That increasing rural to urban migration should speed up the time to crossover is perhaps intuitively understandable; that it should also reduce the urban population growth rate, however, is not. Yet Table 2.1 suggests this conclusion. For example, introducing the Soviet Union's migration rates into India's growth regime results in a lower rate of urban population growth. A similar reduction occurs when the urban growth effects of the growth regime in Table 2.1B.2 are contrasted with those of Table 2.1B.1. What is the cause of this counterintuitive pattern of evolution?

The fixed-rate projection model used to generate the results set out in Table 2.1 defines the urban population growth rate $r_u(t)$ to be the sum of a fixed rate of natural increase, $b_u - d_u$, and a changing rate of urban net inmigration, $m_u(t)$. Given that urban net inmigration reflects the difference between rural and urban outmigration flows, for a national population that is 100 U(t)% urban, we have that

$$m_u(t)U(t) = o_v[1 - U(t)] - o_uU(t), \qquad [2.14]$$

whence, as in 2.9 and 2.10,

$$m_u(t) = o_v \left[\frac{1 - U(t)}{U(t)}\right] - o_u \qquad [2.15]$$

with $m_u(t) > 0$ if $[1 - U(t)]/U(t) > o_u/o_v$.

Because o_v and o_u are fixed by assumption, if U(t) increases with t, then $m_u(t)$ must decrease over time. Hence $r_u(t)$ must decrease also, and so must the fraction of urban growth due to migration (recall that the rate of natural increase is fixed). And because, in our illustrations, increasing o_v increases the urban fraction more than proportionately, $m_u(t)$ and $r_u(t)$ must take on lower values than before.

A projection model that guarantees an ultimately declining fraction of urban growth due to migration is of limited value for answering the question of whether natural increase or net migration is the principal source of urban population growth. It appears that a more realistic model is needed, one that allows the urban natural increase rate to change over time along with the rate of urban net inmigration. The simplest way to introduce such realism is to disaggregate the population by age, as we shall do in Chapters 3 and 4.

The Matrix Projection Model and Stable Growth

Matrix algebra provides a compact and useful means for studying the demographic evolution of multiple interacting populations (Rogers, 1968). Matrix notation makes the projection process more transparent, and matrix theory brings to demographic analysis results that have direct application to population questions. Expressing the population projection process in matrix form also leads to the derivation of results that would be virtually impossible to establish otherwise.

The reader should confirm that the simple biregional projection of the Soviet Union's urban and rural populations described in equations 2.6 and 2.7, respectively, may be expressed in matrix form as

$$\begin{bmatrix} 139 \\ 105 \end{bmatrix} = \begin{bmatrix} 0.998 & 0.035 \\ 0.011 & 0.975 \end{bmatrix} \begin{bmatrix} 136 \\ 106 \end{bmatrix}$$

or, more generally, as

$$\begin{bmatrix} P_u(t+1) \\ P_v(t+1) \end{bmatrix} = \begin{bmatrix} g_{uu} & g_{vu} \\ g_{uv} & g_{vv} \end{bmatrix} \begin{bmatrix} P_u(t) \\ P_v(t) \end{bmatrix} \qquad [2.16]$$

or

$$P(t+1) = G\,P(t) \qquad [2.17]$$

Recalling 2.6 and 2.7, it is evident that

$$\begin{aligned} g_{uu} &= 1 + b_u - d_u - o_u \\ g_{vu} &= o_v \\ g_{vv} &= 1 + b_v - d_v - o_v \\ g_{uv} &= o_u \end{aligned} \qquad [2.18]$$

If these rates are fixed over time, then

$$P(t+n) = G\,P(t+n-1) = G^2\,P(t+n-2) = \ldots = G^n\,P(t) \qquad [2.19]$$

The Soviet Union example set out in Figure 2.1B illustrates such a projection for n = 30 years. The urbanization level is projected to increase from its initial

$$U(1970) = \frac{136}{136 + 106} = 0.56$$

to

$$U(2000) = \frac{226.6}{226.6 + 93.4} = 0.71$$

at which time the urban growth rate should stand at 1.2% per year, about half the rate that was exhibited in 1970.

The results in Table 2.1 indicate that the Soviet Union's urbanization level is expected to grow until it stabilizes at 75% of the total national population. At that point, the population may be said to be experiencing stable growth—a condition characterized by an unchanging "intrinsic" rate of growth (0.09% per year in this example) and a fixed distribution of the population across regions (75% and 25% in this example).

A more transparent picture of the evolution to stable growth may be obtained by focusing on a simplified hypothetical growth process, one that is unencumbered by the clutter of observed data. Imagine an urban population of 24 million that each year sends a fourth of its population to rural areas and receives, in exchange, one-half of the rural population, which initially is also taken to stand at 24 million persons. Assume that a zero population growth regime prevails, such that the annual increment due to births, in each region, is exactly offset by the annual decrement due to deaths. Then we have that

$$G = \begin{bmatrix} 3/4 & 1/2 \\ 1/4 & 1/2 \end{bmatrix}$$

$$P(t) = \begin{bmatrix} 24 \\ 24 \end{bmatrix}$$

and the projection process defined by 2.17 is

$$\begin{bmatrix} 30 \\ 18 \end{bmatrix} = \begin{bmatrix} 3/4 & 1/2 \\ 1/4 & 1/2 \end{bmatrix} \begin{bmatrix} 24 \\ 24 \end{bmatrix}$$

$$\begin{bmatrix} 31\ 1/2 \\ 16\ 1/2 \end{bmatrix} = \begin{bmatrix} 3/4 & 1/2 \\ 1/4 & 1/2 \end{bmatrix} \begin{bmatrix} 30 \\ 18 \end{bmatrix} = \begin{bmatrix} 11/16 & 5/8 \\ 5/16 & 3/8 \end{bmatrix} \begin{bmatrix} 24 \\ 24 \end{bmatrix}$$

$$\begin{bmatrix} 31\ 7/8 \\ 16\ 1/8 \end{bmatrix} = \begin{bmatrix} 3/4 & 1/2 \\ 1/4 & 1/2 \end{bmatrix} \begin{bmatrix} 31\ 1/2 \\ 16\ 1/2 \end{bmatrix} = \begin{bmatrix} 43/64 & 21/32 \\ 21/64 & 11/32 \end{bmatrix} \begin{bmatrix} 24 \\ 24 \end{bmatrix}$$

$$\vdots \qquad \vdots \qquad \vdots$$

$$\begin{bmatrix} 32 \\ 16 \end{bmatrix} = \begin{bmatrix} 3/4 & 1/2 \\ 1/4 & 1/2 \end{bmatrix} \begin{bmatrix} 32 \\ 16 \end{bmatrix} = \begin{bmatrix} 2/3 & 2/3 \\ 1/3 & 1/3 \end{bmatrix} \begin{bmatrix} 24 \\ 24 \end{bmatrix}$$

Note that once the initial urbanization level of 1/2 grows to 2/3, it remains at that level forever. The population has achieved stable growth: Each of its subgroups is increasing exponentially and at the same rate. Its urban and rural growth rates both are zero, and its stable distribution is forever fixed in the proportions 2/3 and 1/3. These two fundamental attributes of the process of projection to stability are augmented by a third: The independence of the stable growth results from the starting population distribution—a property of the process called "ergodicity."

That the stable or intrinsic growth rate and corresponding stable distribution are independent of the starting population distribution and depend only on the growth regime defined by the projection matrix, *G* may be illustrated by applying the same matrix to a different initial population distribution. For example the reader should confirm that

$$\begin{bmatrix} 34 \\ 14 \end{bmatrix} = \begin{bmatrix} 3/4 & 1/2 \\ 1/4 & 1/2 \end{bmatrix} \begin{bmatrix} 40 \\ 8 \end{bmatrix}$$

converges to the same stable state as was obtained before, and that the projection matrix

$$\begin{bmatrix} 5/6 & 1/4 \\ 1/4 & 5/6 \end{bmatrix}$$

ultimately brings about a level of urbanization with half of the national population living in rural areas and growing at 25/3 = 8.3% per annum.

Population Redistribution in Belgium

In its new 1970 constitution, Belgium was officially divided into three principal regions: Brussels, Flanders, and Wallonia. Table 2.2 sets out data on population stocks, births, deaths, and migration for this three-region system, and Table 2.3 transforms these data into rates. Expanding the matrix model defined in equation 2.17, we obtain the following projection of the end-of-year population in 1970:

$$\begin{bmatrix} 1{,}073{,}998 \\ 5{,}413{,}782 \\ 3{,}157{,}399 \end{bmatrix} = \begin{bmatrix} 0.969497 & 0.002615 & 0.004221 \\ 0.017749 & 1.000175 & 0.002383 \\ 0.012907 & 0.001435 & 0.993583 \end{bmatrix} \begin{bmatrix} 1{,}079{,}520 \\ 5{,}386{,}158 \\ 3{,}155{,}988 \end{bmatrix}$$

Differences between the projected and the observed end-of-year populations may be attributed to inaccurate reporting of event and flow data and to the omission of the effects of international migration. International migration may be introduced into the model by adding the effects of net immigration in the diagonal of the projection matrix. This would result in revised diagonal values of 0.970552, 1.000695, and 0.994162, respectively. For simplicity, however, we shall ignore the impacts of such international migration.

Raising the projection matrix to successively higher powers, we find that at stability the Belgian population would be increasing at an annual compounded rate of 0.3% a year. Its stable distribution at that point would allocate 8.2% of the national population to the Brussels region, 69.9% to Flanders, and 21.9% to Wallonia (Willekens and Philipov, 1981). To confirm these results we observe that the following equality is satisfied:

$$\begin{bmatrix} 8{,}237 \\ 70{,}107 \\ 21{,}957 \end{bmatrix} = \begin{bmatrix} 0.969497 & 0.002615 & 0.004221 \\ 0.017749 & 1.000175 & 0.002383 \\ 0.012907 & 0.001435 & 0.993583 \end{bmatrix} \begin{bmatrix} 8{,}212 \\ 69{,}897 \\ 21{,}891 \end{bmatrix}$$

$$= (1.00301) \begin{bmatrix} 8{,}212 \\ 69{,}897 \\ 21{,}891 \end{bmatrix}$$

TABLE 2.2 Components of Population Change in a Three-Region System: Belgium, 1970

	Internal Migration				Natural Change		International Migration		Population		
Region *to*	*Brussels*	*Flanders*	*Wallonia*	*Total*	*Births*	*Deaths*	*In*	*Out*	*31 Dec '69*	*1 July '70*	*31 Dec '70*
from	(1)	(2)	(3)	(4)	(5)	(6)	(7)	(8)	(9)	(10)	(11)
Brussels	0	19,160	13,934	33,094	14,694	14,529	19,086	17,947	1,079,520	1,077,328	1,075,136
Flanders	14,085	0	7,728	21,813	82,586	59,832	20,189	17,387	5,386,158	5,401,370	5,416,583
Wallonia	13,321	7,522	0	20,843	44,888	44,299	22,866	21,037	3,155,988	3,157,606	3,159,225
Total	27,406	26,682	21,662	75,750	142,168	118,660	62,141	56,371	9,621,666	9,636,304	9,650,944

SOURCE: Willekens (1979).

TABLE 2.3 Rates of Population Change in a Three-Region System: Belgium, 1970

		Internal Migration				Natural Change		International Migration	
Region	*to*	*Brussels*	*Flanders*	*Wallonia*	*Total*	*Births*	*Deaths*	*In*	*Out*
from									
Brussels		0	0.017749	0.012907	0.030656	0.013612	0.013459	0.017680	0.016625
Flanders		0.002615	0	0.001435	0.004050	0.015333	0.011108	0.003748	0.003228
Wallonia		0.004221	0.002383	0	0.006604	0.014223	0.014036	0.007245	0.006666

SOURCE: Willekens (1979).

A population of 100,000 people, allocated among the three regions according to the stable distribution, would increase by 0.3% in each region every year.

A comparison of the stable regional allocation of the national population to the one observed in 1970 shows Flanders gaining in relative population size. The reader may wish to examine the sources of this population growth and to identify the relative contribution made to it by internal migration.

The division of a national population into subnational regional populations is but one of many alternative disaggregations frequently adopted in formal population analyses and projections. Probably the most common is a division of the population into age groups. We take up this topic in the next chapter.

3. UNIREGIONAL POPULATION DYNAMICS: AGE WITHOUT LOCATION

Inventories and projections of human populations are an important component of most socioeconomic planning efforts. Populations are the clients whose welfare the planning is supposed to improve; they also are a primary resource used in the production of goods and services that lead to higher levels of welfare; and they consume resources that might be more profitably used elsewhere.

But welfare levels, participation patterns in production, and consumption behavior all vary with age. For example, demands for goods and services often increase or decline in rough proportion to the pattern of population change in certain age groups. The need for elementary schools and elementary school teachers falls with declines in the number of children, as does the market for baby food. Demand for police protection and prisons increases with the growth of young adults in the ages of peak criminal activity, and health care requirements and the job market for nurses grow with a rise in the number of persons in the pensionable age groups.

A disaggregation by age, then, is introduced into subnational population projections because forecasts of these population subgroups are important in their own right. But disaggregation also may be advocated because demographic rates are more stable and more meaningful when they refer to relatively homogeneous groups. For example, schedules of age-specific mortality rates normally show a moderately high death rate immediately after birth, followed by a drop to a minimum between ages 10 to 15, then a slow and gradual increase until about age 50, and thereafter a rise at an increasing pace until the last years of life. Fertility rates generally start to assume positive values at age 15, rise to attain a maximum somewhere between ages 20 and 30, and decline to zero once again at an age close to 50. Rates of migration start out with moderately high values for infants, drop to a low point at about age 16, turn sharply upward to a peak near ages 20 to 22, and then decline regularly thereafter until the onset of the principal ages of retirement, at which point a slight hump or an upwardly sloping curve may be evidenced.

In introducing age to the subnational population projection, we begin with the simplest case—a population exposed to fixed birth and death rates and zero migration.

The Uniregional Projection Model

Imagine a regional population, undisturbed by migration, that has been disaggregated into five-year age groups and whose evolution is to be projected forward over a unit time interval of five years. If the number of individuals aged x to x + 4 at last birthday is denoted by P(x), then the number five years later will be s(x)P(x), where s(x) is the fraction surviving from one age group to the next. Normally s(x) comes from a life table that describes the mortality pattern of the population being projected.

The calculation of the survivors of an initial population distribution at time t, set out as the vector *P*(t), may be conveniently expressed in matrix form as *S P*(t), where all elements of the matrix *S* are zero except those along the subdiagonal. Thus we have that

$$\left.\begin{aligned} P(t+1) &= S\,P(t) \\ P(t+2) &= S\,P(t+1) \\ &\;\;\vdots \\ P(t+n) &= S\,P(t+n-1) \end{aligned}\right\} \qquad [3.1]$$

or

$$P(t+n) = S^{n}\,P(t) \qquad [3.2]$$

where

$$P(t) = \begin{bmatrix} P(0;t) \\ P(5;t) \\ \cdot \\ \cdot \\ \cdot \\ \cdot \end{bmatrix} \quad \text{and } S = \begin{bmatrix} 0 & 0 & 0 & \cdots \\ s(0) & 0 & 0 & \cdots \\ 0 & s(5) & 0 & \\ \cdot & \cdot & \cdot & \\ \cdot & \cdot & \cdot & \\ \cdot & \cdot & \cdot & \end{bmatrix} \qquad [3.3]$$

such that the jth element of the ith row is $s_{ij}(x)$ for $j = i - 1 = 1, 2, 3, \ldots$ and 0, otherwise.

Repeated multiplication of $P(t)$ by S will ultimately produce a vector of zeros because additions due to births are not included. To incorporate the impact of fertility, we let b(x) denote the average number of babies born during the unit age-time interval and alive at the end of that interval, per person aged x to x + 4 at last birthday at the start of the interval. That age group's contribution to the first age group, then, is b(x)P(x; t), and summing over all ages of childbearing (here taken to be from ages α to β), gives

$$P(0; t+1) = \sum_{x=\alpha-5}^{\beta-5} b(x)\, P(x; t) \qquad [3.4]$$

Defining a matrix B that has zeros everywhere except in the first row allows us to express the population projection model as

$$P(t+1) = (S + B)\, P(t) \qquad [3.5]$$

or, more simply,

$$P(t+1) = G\, P(t) \qquad [3.6]$$

where

$$G = \begin{bmatrix} 0 & 0 & b(10) & b(15) & \cdots \\ s(0) & 0 & 0 & \cdots & \\ 0 & s(5) & 0 & & \\ \cdot & \cdot & \cdot & & \\ \cdot & \cdot & \cdot & & \\ \cdot & \cdot & \cdot & & \\ \cdot & \cdot & \cdot & & \end{bmatrix} \qquad [3.7]$$

Note that equation 3.6 has the same matrix expression as equation 2.15. Only the contents of the matrix and the vectors have been altered.

The projection in 3.6 could be carried out first for females and then for males, using for each the appropriate life table and age-specific rates of motherhood and fatherhood. However, dealing with the two sexes separately can introduce discrepancies in long-term projections, and demographers therefore prefer to attribute the births of both boys and girls to the mother—in what is called the female dominant model. A projection matrix that incorporates both sexes may be readily constructed on such a model, with equation 3.6 now incorporating a much larger matrix G and longer vector $P(t)$. Supposing that the upper left quarter of G projects females and the lower two quarters males, we have

$$\begin{bmatrix} P_f(t+1) \\ P_m(t+1) \end{bmatrix} = \begin{bmatrix} S_f + B_f & 0 \\ kB_m & S_m \end{bmatrix} \begin{bmatrix} P_f(t) \\ P_m(t) \end{bmatrix} \qquad [3.8]$$

where k is the product of (1) the ratio of male to female births, and (2) the ratio of male to female stationary life table populations in the first age group. Note that the births of baby girls and baby boys, generated by the fertility elements of B_f and B_m in the multiplication process, are solely a function of the female population $P_f(t)$.

Once the population projection process has been expressed in matrix form, we can draw on the theory of matrices with nonnegative elements to study its properties (Rogers, 1975). This theory informs us that, except for a set of odd fertility patterns that are not observed in human populations, repeated powering of the matrix G ultimately will produce a matrix with only positive elements. And eventually each element of the matrix with the higher power will be proportional to the corresponding element of the matrix with the lower power; that is,

$$G^{h+1} \doteq \lambda \; G^h \qquad [3.9]$$

as was the case in Chapter 2. Once again λ may be interpreted as the growth ratio, in this case per five-year period, and the stable intrinsic annual rate of growth is given by 0.2 ln λ, since $\lambda = e^{5r}$.

Consider the following simple numerical example. Imagine a population disaggregated into four age groups: 0-14, 15-29, 30-44, and 45 years and over. Suppose that 216 thousand individuals are to be found in each age group and that the fractions surviving, s(x), are all equal to

5/6. Assume that the fertility elements, b(x), are 0, 3/4, 5/6, and 1/4, respectively. Then the projection defined in 3.5 yields

$$\begin{bmatrix} 396 \\ 180 \\ 180 \\ 180 \end{bmatrix} = \begin{bmatrix} 0 & 3/4 & 5/6 & 1/4 \\ 5/6 & 0 & 0 & 0 \\ 0 & 5/6 & 0 & 0 \\ 0 & 0 & 5/6 & 0 \end{bmatrix} \begin{bmatrix} 216 \\ 216 \\ 216 \\ 216 \end{bmatrix}$$

because

$$396 = 0\,(216) + 3/4\,(216) + 5/6\,(216) + 1/4\,(216)$$

and

$$180 = 5/6\,(216)$$

for the remaining three age groups.

Note that a more realistic projection would replace the zero in the lower right corner of the growth matrix by some positive value in order to allow a fraction of those 45 and over to survive the unit time interval and to remain once again in the same 45-and-over age group. In most empirical projections, however, the open-ended last age group is at least 85 years and over and the matter is usually resolved by inflating the next-to-last age group's fraction surviving in order to account for individuals that remain in the last age group, that is, s(80) is inflated and therefore may exceed unity.

The reader should confirm that the above matrix projection process ultimately stabilizes and exhibits both an unchanging growth ratio of λ = 1.1209 and a fixed age composition of (0.3694; 0.2746; 0.2042; 0.1518).

The Life Table

Vital statistics and censuses provide the necessary data for the calculation of age-specific birth and death rates, which may be used to answer questions such as: What is the current rate at which 40-year-old males are dying from heart disease or at which 30-year-old women are bearing their second child? But many of the more interesting questions regarding mortality and fertility patterns are phrased in terms of probabilities; for example: What is the current probability that a man aged 40 will outlive his 38-year-old wife, or that she will bear a third child before her forty-fifth birthday?

Demographers normally estimate probabilities from observed rates by developing a life table. Such tables describe the evolution of a hypothetical cohort of babies born at a given moment and exposed to an unchanging age-specific schedule of rates. For this cohort of babies, they exhibit a number of probabilities for changes of state, such as dying, and develop the corresponding expectations of life spent in different states at various ages. Life tables that deal with age intervals of a year are commonly referred to as *complete* life tables, whereas those using longer age intervals are called *abridged* life tables. However, we shall ignore this somewhat spurious distinction and for expositional convenience will focus only on age intervals five years wide.

The simplest life tables recognize only one category of decrement, death, and their construction is normally initiated by estimating a set of age-specific probabilities of dying within each interval of age, q(x) say, from observed data on age-specific death rates, M(x) say. The conventional calculation assumes that deaths are uniformly distributed over time and over the ages within the age interval. If D(x) denotes the number of deaths registered during a calendar year among people aged x to x + 4 at last birthday, we may assert that people in that age interval died at the rate of D(x) in each of the 5 years and that (5/2) D(x) represents the number of people who were alive at the beginning of the unit interval and who died by the time it was half over. Thus if P(x) denotes the population at the mid-point of the interval, then 5D(x) is the number of deaths over the 5-year period, P(x) + 5/2 D(x) is the population exposed on average to the possibility of dying, and

$$q(x) = \frac{5D(x)}{P(x) + 5/2\, D(x)} \qquad [3.10]$$

Dividing both the numerator and denominator by P(x), we have that

$$q(x) = \frac{5M(x)}{1 + 5/2\, M(x)} \qquad [3.11]$$

or, alternatively,

$$p(x) = 1 - q(x) = [1 + 5/2\, M(x)]^{-1}\, [1 - 5/2\, M(x)] \qquad [3.12]$$

where M(x) = D(x)/P(x), and p(x) is the age-specific probability of surviving from exact age x to exact age x + 5.

The annual death rate among 40 to 44-year-olds in India in 1970 has been estimated to be 0.006797 (Appendix A). Thus the associated probability of dying between exact age 40 and exact age 45 is

$$q(40) = \frac{5(0.006797)}{1 + 5/2\,(0.006797)} = 0.033418$$

and the corresponding probability of surviving is

$$p(40) = 1 - q(40) = 0.966582$$

These values may be found in Columns 2 and 3 of Table 3.1, which presents a life table for India's total (male plus female) population in 1970. Notice that the corresponding death rate appears in Column 7. This is because the above method of calculating q(x) implies an equality between the observed population's death rate, M(x), and its life table population's counterpart, m(x).

All of the columns in a life table originate from a set of probabilities of dying at each age, q(x). By applying, in sequence, a particular set of such probabilities to a cohort of arbitrary size, commonly taken to be 100,000 babies and denoted by $\ell(0)$, we can observe the diminution of this cohort (referred to by demographers as the life table's *radix*) as the life table population ages. For example, the number of survivors at each exact age is given by

$$\ell(x + 5) = [1 - q(x)]\,\ell(x) \qquad [3.13]$$

This implies that deaths can be obtained by subtraction as

$$d(x) = \ell(x) - \ell(x + 5) \qquad [3.14]$$

The number of years lived by the life table cohort in each age group is denoted by L(x). Generally, this measure is calculated by assuming that the curve of survivors, $\ell(x)$, declines linearly from one age to the next. Thus, over a five-year interval, the $\ell(x + 5)$ survivors out of the initial $\ell(x)$ individuals lived a full 5 years, whereas those who died in the interval lived, on average, two-and-one-half years. Hence,

$$\begin{aligned} L(x) &= 5\ell(x + 5) + 5/2\,d(x) \qquad [3.15] \\ &= \ell(x + 5) + 5/2\,[\ell(x) - \ell(x + 5)] \\ &= 5/2\,[\ell(x) + \ell(x + 5)] \end{aligned}$$

TABLE 3.1 Uniregional Life Table: India, 1970

Age, x	p(x)	q(x)	ℓ(x)	d(x)	L(x)	m(x)	s(x)	T(x)	e(x)
(1)	(2)	(3)	(4)	(5)	(6)	(7)	(8)	(9)	(10)
0	0.761213	0.238787	100,000	23,879	440,303	0.054232	0.852517	4,885,568	48.86
5	0.972463	0.027537	76,121	2,096	375,366	0.005584	0.980073	4,445,265	58.40
10	0.987898	0.012102	74,025	896	367,886	0.002435	0.987558	4,069,899	54.98
15	0.987213	0.012787	73,129	935	363,309	0.002574	0.984007	3,702,013	50.62
20	0.980761	0.019239	72,194	1,389	357,498	0.003885	0.980883	3,338,704	46.25
25	0.981007	0.018993	70,805	1,345	350,664	0.003835	0.979759	2,981,206	42.10
30	0.978487	0.021513	69,460	1,494	343,566	0.004349	0.976260	2,630,542	37.87
35	0.973983	0.026017	67,966	1,768	335,410	0.005272	0.970332	2,286,976	33.65
40	0.966582	0.033418	66,198	2,212	325,459	0.006797	0.960005	1,951,566	29.48
45	0.953202	0.046798	63,986	2,994	312,442	0.009584	0.941267	1,626,107	25.41
50	0.928746	0.071254	60,991	4,346	294,092	0.014777	0.915744	1,313,665	21.54
55	0.901744	0.098256	56,645	5,566	269,312	0.020667	0.871674	1,019,573	18.00
60	0.838328	0.161672	51,080	8,258	234,753	0.035178	0.814747	750,260	14.69
65	0.786618	0.213382	42,821	9,137	191,264	0.047773	1.695266*	515,508	12.04
70+	0.	1.000000	33,684	33,684	324,244	0.103885	0.	324,244	9.63

SOURCES: Rogers (1982b) and Appendix A.

*This s(x) exceeds unity because it refers to survivorship into an open-ended age interval. Because not all members in that interval die over a period of five years, a "correction" must be incorporated into the value of s(x).

For example, in Column 4 of Table 3.1 we see that 63,986 of 66,198 persons at age 40 survived to age 45. If the 2,212 deaths were distributed uniformly over the age interval, then the 2,212 individuals who died lived an average of two-and-one-half years each, or 5/2 (2,212) = 5,530 person-years. The 63,986 who survived lived a full 5 years each, or 5 (63,986) = 319,930 person-years. Hence the total number of person-years lived by the cohort during the five-year interval between ages 40 and 45, was

$$\begin{aligned} L(40) &= 5\,(63{,}986) + 5/2\,(2{,}212) \\ &= 319{,}930 + 5{,}530 \\ &= 325{,}460 \end{aligned}$$

The slight discrepancy from the number set out in Table 3.1 is due to rounding.

An alternative interpretation of $\ell(x)$ and $L(x)$ is also useful. Instead of thinking of these measures as referring to the descendants of a cohort of 100,000 babies, we may view them instead as describing a stationary (zero growth) life table population that is fixed in size as well as in age composition. This population experiences 100,000 births and deaths each year, and includes $\ell(x)$ persons at exact age x and $L(x)$ individuals in the age group x to x + 4. This latter interpretation is particularly useful in understanding how to define $s(x)$, the fraction of the population in each age group that survives to the next age group. For if $L(x + 5)$ denotes the survivors of those previously included in $L(x)$, then clearly

$$s(x) = \frac{L(x+5)}{L(x)} = L(x+5)\,L(x)^{-1} \qquad [3.16]$$

Another life table statistic is the age-specific life table death rate, $m(x)$, which is defined as

$$m(x) = \frac{d(x)}{L(x)} \qquad [3.17]$$

If, as in Table 3.1, this life table death rate is equal to its counterpart in the observed population, then

$$L(x) = \frac{d(x)}{M(x)} \qquad [3.18]$$

The last two columns of the life table describe the total number of person-years lived (or total life table population) beyond each age, T(x), and the average expectation of remaining life at each age, e(x). The former is simply the sum of the L(x) values after a give age—a quantity that if divided by the $\ell(x)$ at that age defines the latter statistic. Thus

$$T(x) = \sum_{y=x}^{\infty} L(y) \qquad [3.19]$$

and

$$e(x) = \frac{T(x)}{\ell(x)} \qquad [3.20]$$

The value e(0) is the expectation of life at birth. For India, on 1970 rates, it was 48.86 years according to Table 3.1.

The terminal age group in a life table is open-ended, z years and over. For this age group, q(z) is set equal to unity, and the statistics M(z), $\ell(z)$, T(z), and e(z) all refer to the interval age z and over. Thus it is necessary to modify equation 3.15 in order to calculate L(x). Setting x = z in 3.18 gives

$$L(z) = \frac{d(z)}{M(z)} \qquad [3.21]$$

and, because each of the $\ell(z)$ individuals will ultimately die, $\ell(z) = d(z)$, whence

$$L(z) = \frac{\ell(z)}{M(z)} = M(z)^{-1}\ell(z) \qquad [3.22]$$

By definition,

$$T(z) = L(z) \qquad [3.23]$$

consequently,

$$e(z) = \frac{T(z)}{\ell(z)} = \frac{L(z)}{\ell(z)} \qquad [3.24]$$

TABLE 3.2 National Population Distribution, Death Rates, and Birth Rates: Soviet Union, 1970

Age *x*	*Population (in thousands)*	*Death Rate* *M(x)*	*Birth Rate* *F(x)*
0	20,533	0.0070	
5	24,503	0.0007	
10	25,017	0.0006	
15	22,023	0.0010	0.0152
20	17,124	0.0016	0.0828
25	13,785	0.0022	0.0662
30	21,168	0.0028	0.0455
35	16,612	0.0038	0.0251
40	19,024	0.0048	0.0084
45	12,269	0.0061	0.0018
50	9,088	0.0088	
55	12,027	0.0119	
60	10,348	0.0182	
65	7,267	0.0278	
70+	10,932	0.0766	
Total	241,720	0.0083	0.1748

SOURCE: Rogers (1978a).

Urbanization in the Soviet Union Revisited

In Chapter 1 we considered a uniregional urbanization projection model that ignored age. The projected urban population was obtained as a residual—the difference between the projected national and rural population totals. An analogous approach may be used to obtain the projected urban population disaggregated by age. One first projects the national and rural age distributions and then takes the urban population to be their difference.

Consider the demographic data for the Soviet Union that are set out in Table 3.2. A uniregional life table for the Soviet Union gives an expectation of life at birth of 70.04 years and a life table population of L(0) = 491,405 individuals in the first five-year age group. Ratios of consecutive age-specific life table populations define the s(x) values. How does one obtain the fertility elements b(x) from the observed birth rates F(x)?

Usually the annual birth rate, F(x), is applied to the arithmetic mean of the initial and final populations aged x to x + 4 years at last birthday:

$$\frac{P(x;t) + P(x;t+1)}{2} = 1/2\ [P(x;t) + s(x-5)\ P(x-5;t)] \qquad [3.25]$$

and, because this population is exposed to the fertility regime over a time interval of five years, we multiply equation 3.25 by 5. Thus this age group's contribution to the total number of births during the five years is

$$5/2\,[P(x;t) + s(x-5)\,P(x-5;t)]\,F(x) \qquad [3.26]$$

Adding this quantity over all of the childbearing age groups, starting with α and ending with $\beta - 5$, yields (Keyfitz, 1968, p. 30):

$$5/2 \sum_{x=\alpha}^{\beta-5} [P(x;t) + s(x-5)\,P(x-5;t)]\,F(x)$$

$$= 5/2 \sum_{x=\alpha-5}^{\beta-5} [F(x) + s(x)F(x+5)]\,P(x;t) \qquad [3.27]$$

Equation 3.27 describes the total number of births that are expected during the five-year time interval. But what is needed is the total number of babies that survive to the start of the next time interval. Hence 3.27 should be multiplied by the survival factor $L(0)/5\ell(0)$. Thus we have that

$$P(0;t+1) = \sum_{x=\alpha-5}^{\beta-5} \frac{L(0)}{2\ell(0)} [F(x) + s(x)\,F(x+5)]\,P(x;t)$$

$$= \sum_{x=\alpha-5}^{\beta-5} b(x)\,P(x;t) \qquad [3.28]$$

where

$$b(x) = 1/2\,L(0)\,\ell(0)^{-1}\,[F(x) + F(x+5)\,s(x)] \qquad [3.29]$$

If $\alpha = 15$, the assumed age when women can start bearing children, we obtain the projection matrix:

$$G = \begin{bmatrix} 0 & 0 & \frac{L(0)}{2\ell(0)}\left(\frac{L(15)}{L(10)} F(15)\right) & \frac{L(0)}{2\ell(0)}\left(F(15) + \frac{L(20)}{L(15)} F(20)\right) & \dots & 0 \\ \frac{L(5)}{L(0)} & 0 & 0 & 0 & \dots & 0 \\ 0 & \frac{L(10)}{L(5)} & 0 & 0 & \dots & 0 \\ \vdots & \vdots & \ddots & & & \vdots \\ 0 & 0 & & & & 0 \end{bmatrix} \qquad [3.30]$$

The contribution made to the first age group in the Soviet Union at time t + 1 by surviving children of 20- to 24-year-old parents at time t is

$$b(20) = 1/2\,(4.9141)\,[F(20) + 0.9904\,F(25)] \qquad [3.31]$$

into which we may substitute F(20) = 0.0828 and F(25) = 0.0662 to find b(20) = 0.3645.

A uniregional age-disaggregated projection of the Soviet Union's 1970 population to the year 2000, using the data set out in Table 3.2 and the matrix model defined by equations 3.6 and 3.30, gives rise to the age distribution presented in Column 2 of Table 3.3. The 1970 total of 241.7 million persons is expected to grow to 322.9 million by the year 2000, and the average annual rate of growth during the 1995-2000 interval should be about 1.1% per annum.

Repeating the above uniregional projection exercise with the Soviet Union's 1970 rural population, using data set out in Rogers and Philipov (1980) and treating net rural outmigration as another source of "mortality" gives rise to the rural population totals presented in Column 4 of Table 3.3. Subtracting these totals from those set out in Column 2 gives the figures for the urban population presented in Column 3. The reader should compare the results of these projections of the Soviet Union's urbanization with those presented earlier in Chapters 1 and 2.

Zero Population Growth

Modern projections of world population have adopted the component approach in two distinct senses of the word. First, they have

TABLE 3.3 Uniregional Cohort-Survival Projections of Total, Rural, and (by subtraction) Urban Populations of the Soviet Union to the Year 2000

Age	*Population (in thousands)**			*Age Composition**		
x	*Total*	*Urban*	*Rural*	*Total*	*Urban*	*Rural*
(1)	*(2)*	*(3)*	*(4)*	*(5)*	*(6)*	*(7)*
0	26,181	20,574	5,607	0.0811	0.0800	0.0855
5	25,563	19,569	5,994	0.0792	0.0761	0.0914
10	25,931	19,325	6,606	0.0803	0.0751	0.1008
15	25,379	19,299	6,080	0.0786	0.0750	0.0927
20	23,226	19,193	4,033	0.0719	0.0746	0.0615
25	20,828	18,048	2,780	0.0645	0.0701	0.0424
30	19,421	16,587	2,834	0.0602	0.0645	0.0432
35	23,243	20,067	3,176	0.0720	0.0780	0.0484
40	23,307	20,323	2,984	0.0722	0.0790	0.0455
45	20,050	17,990	2,060	0.0621	0.0699	0.0314
50	15,120	13,271	1,849	0.0468	0.0516	0.0282
55	11,673	9,071	2,602	0.0362	0.0353	0.0397
60	16,847	12,422	4,425	0.0522	0.0483	0.0675
65	11,993	8,154	3,839	0.0371	0.0317	0.0586
70+	34,092	23,397	10,695	0.1056	0.0909	0.1631
Total	322,855	257,290	65,563	1.0000	1.0000	1.0000
Share	1.0000	0.7969	0.2031			
Annual growth rate	0.0108	0.0157	−0.0133			

SOURCE: Rogers and Philipov (1980).
*Slight differences are due to independent rounding.

identified the separate contributions of age-specific rates of fertility and mortality to changing patterns of population growth and composition; and, second, they have focused attention on the diverse patterns of growth and change exhibited by individual national and regional populations, aggregating these separate totals to obtain global projections. Such efforts have been made possible by recent advances in demographic knowledge, data, and computational methods and facilities.

One of the early contributions to modern global population projections was the set prepared by Frejka (1973). What distinguished his projections from those previously carried out by others was his explicit adoption of a set of assumptions that specified the period of time during which fertility in each country or world region was expected to change from its present level to bare replacement level. It was assumed that fertility would remain constant thereafter.

More recently, Littman and Keyfitz (1977) and the World Bank (1983) have issued global population projections for a large number of individual countries. Such detail enhances the possibilities of critical evaluation and permits the grouping of individual national totals into aggregates on the basis of criteria other than spatial contiguity.

A common feature of most modern global population projections, then, is the notion of replacement level fertility. This is taken to be the fixed level of fertility that ensures ultimate zero population growth, or *stationarity*. A stationary population is a stable population with a zero growth rate. As with all stable populations, the age composition remains constant over time and each population subgroup grows at the same fixed rate—in this instance a zero population growth (ZPG) rate.

Central to the notion of replacement level fertility is the *net reproduction rate*, NRR. This is the average number of children of a single sex (baby girls say) expected to be born in the future to a child of the same sex (baby girl say) just born:

$$\mathrm{NRR} = \sum_{x=0}^{\infty} \frac{L(x)}{\ell(0)} F(x) \qquad [3.32]$$

where L(x) is the life table population aged x to x + 4 years, and F(x) is the annual birth rate of individuals in that age group. Thus NRR may be interpreted as the ratio of the number living in two successive generations that is implied by current rates of birth and death. An absence of migration is assumed, and the population is said to be experiencing replacement level fertility when NRR = 1.

The net reproduction rate confounds the impacts of fertility, F(x), with those of mortality, L(x). A "pure" measure of fertility can be obtained by eliminating the effects of the latter component in equation 3.32. Setting all $L(x) = 5\,\ell(0)$, where 5 is the width of the age interval in years, we have the definition of the *gross reproduction rate*:

$$\mathrm{GRR} = 5 \sum_{x=0}^{\infty} F(x) \qquad [3.33]$$

which is a function only of fertility.

Clearly there are many ways by which a schedule of fertility rates F(x) can be reduced to yield an NRR of unity, given the fixed mortality schedule reflected by the set of L(x) values. Analyses of trajectories toward zero population growth almost always assume that every age-specific birth rate is reduced proportionately. Thus if the original net reproduction rate is above unity, the reduced birth rates are defined by the relation

$$F(x) = \frac{F(x)}{NRR} \qquad [3.34]$$

If the population projection is carried out with the reduced birth rates, F(x), an eventual convergence to zero population growth is ensured if the population is undisturbed by migration.

For example, using the data set out in Appendix A and in Table 3.1, the national net reproduction rate for India can be found to be 1.83. A drop to replacement level fertility would therefore be achieved if each age-specific national birth rate, F(x), were reduced by $100(1 - 1/1.83) = 45.4\%$. Zero population growth would not follow immediately, however, because of the momentum for growth that is embodied in India's age composition.

Because of their much higher fertility, less developed nations have a much younger age composition than developed countries and, therefore, a far greater built-in tendency for further growth. A country with a recent history of high birth rates such as Mexico, for example, exhibits an age composition that has the shape of a pyramid with a broad base at the youngest age groups and a sharp tapering off at the older age groups. On the other hand, a country with a history of low birth rates, such as Sweden, has an age composition that yields an almost rectangular age pyramid (Figure 3.1).

Populations in which children outnumber parents potentially have a larger number of parents in the next generation than today and therefore acquire a built-in momentum for further growth, even if their fertility suddenly drops to bare replacement levels. Thus if fertility levels in developing countries dropped to bare replacement immediately, this would produce zero population growth only after 70 years or more, and the resulting stationary population would be about two-thirds larger than the current one (Figure 3.2). If the drop were to take about 70 years to achieve, the increase would be about 450%. In other words, the momentum with the immediate fertility decline is about $1^2/_3$, and with delayed decline about $5^1/_2$.

Populations in all developed countries have gone through a process of demographic change in which a decline in mortality eventually was followed by a drop in fertility. Demographers refer to this transformation as the demographic transition and associate it with socioeconomic changes that arise during a nation's industrialization and

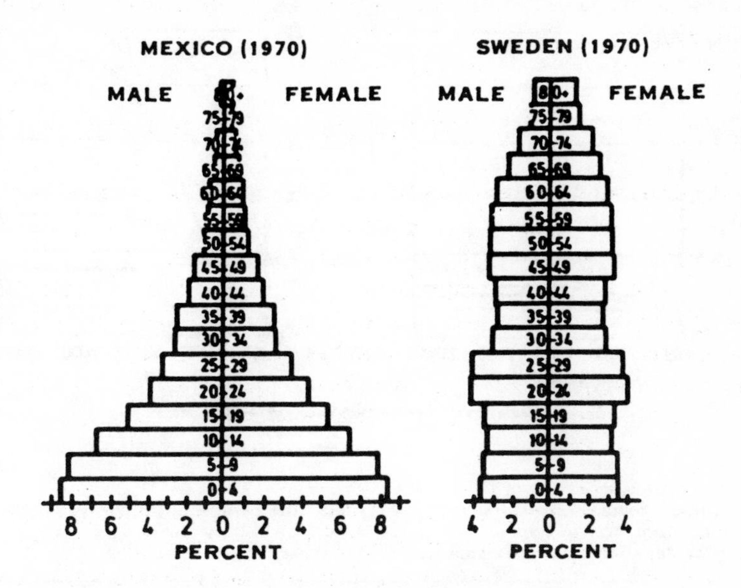

SOURCE: Redrawn from Berelson (1974) with the permission of the Population Council from "World Population: Status Report 1974," by Bernard Berelson, **Reports on Population/Family Planning**, No. 15 (January 1974), pp. 12, 13.

Figure 3.1 Young and Old Population Age Compositions

modernization. Although the progress has been far from uniform, and its linkages with changes in socioeconomic variables have not been clearly identified, the universality of this historical demographic revolution in what today are the developed countries is nevertheless quite impressive.

The National and Global Population Projections of the United Nations

In its 1980 assessment, the Population Division of the United Nations extended for the first time its periodically issued projections of future population to 2025 (United Nations, 1981). This assessment of future prospects, the most important and most widely used source of demographic projections with a global coverage, is built around a central ("medium variant") projection that reflects the evolution predicted by the demographic transition. The projections portray a demographic future that incorporates population stabilization at low levels of fertility

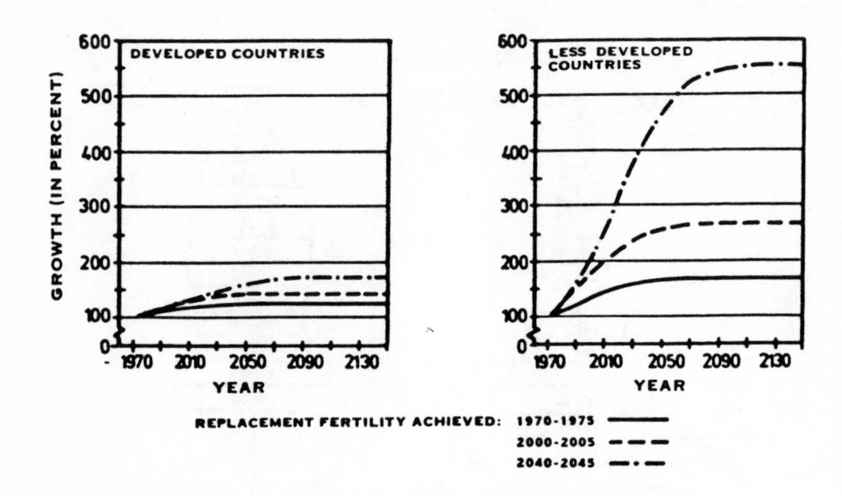

SOURCE: Redrawn from Berelson (1974) with the permission of the Population Council from "World Population: Status Report 1974," by Bernard Berelson, **Reports on Population/Family Planning**, No. 15 (January 1974), pp. 12, 13.

Figure 3.2 Momenta of Population Growth for Developed and Less Developed Countries

and mortality (and international migration), with the only important difference among these projections being the timing of the onset of replacement level fertility. Alternative projections that bracket the medium variant projection are included to illustrate the consequences, to population growth and composition, of particular deviations in the adopted assumptions. Fertility levels in developed countries are assumed to decline or to remain below replacement until around the end of this century, after which time they are assumed to approach replacement level fertility.

> Irrespective of whether the country is developed, with very low fertility (for example, the Federal Republic of Germany or Japan), or developing with high fertility (for example, Bangladesh or the Syrian Arab Republic), it is assumed that fertility will arrive at replacement levels in the not too distant future [United Nations, 1984, pp. 30-32].

Figure 3.3 illustrates past and assumed future trends in fertility, as measured by gross reproduction rates during the period from 1950 to 2025, for some of the larger countries in the world. Notice that once the decline in fertility starts, it is expected to proceed slowly, gain momentum, then ultimately reduce its speed.

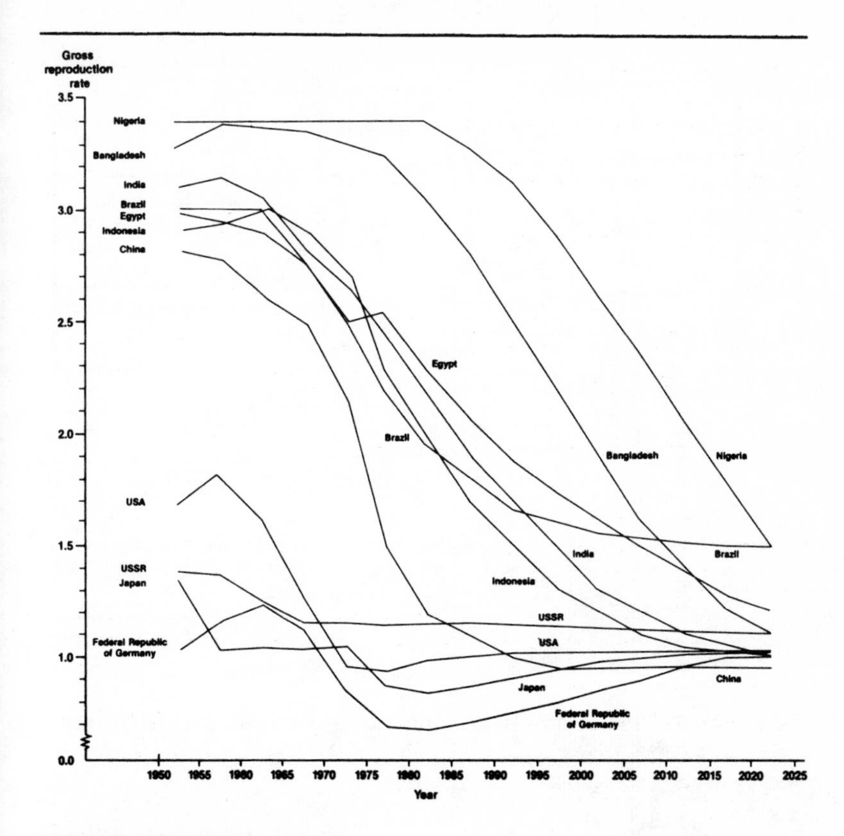

SOURCE: United Nations (1984).

Figure 3.3 Gross Reproduction Rates for Selected Countries, 1950-2025, United Nations Medium Variant

The mortality assumptions in the United Nations projections are expressed in the form of life expectancies at birth and their associated age-sex-specific patterns of survivorship. Model life tables (Coale and Demeny, 1966, 1983) are used to define probabilities of surviving when national life tables are unavailable or unreliable. In general, the projections assume that mortality trends will produce a quinquennial gain of 2.5 years in the expectation of life at birth every five years, until life expectancy reaches 55 years, at which point a slowing of this gain is expected. The maximum level of life expectancy at birth is fixed at 73.5 years for males and 80 years for females, in accordance with current demographic experience in several developed countries.

TABLE 3.4 United Nations Population Estimates and Medium Variant Projections (in millions) for Major Subregions and the World, 1960-2025

			Gross Reproduction Rate		*Expectation of Life at Birth*		*Projected Population*	
	1960	*1980*	*1975-1980*	*2020-2025*	*1975-1980*	*2020-2025*	*2000*	*2025*
World Total	3,037	4,432	1.92	1.15	57.5	70.4	6,119	8,195
More Developed Regions	945	1,131	1.00	1.04	71.9	75.4	1,272	1,377
Less Developed Regions	2,092	3,301	2.27	1.17	55.1	69.6	4,847	6,818
Africa	275	470	3.13	1.49	48.6	67.2	853	1,542
Latin America	216	364	2.24	1.35	62.5	71.8	566	865
Northern America	199	248	0.94	1.02	73.0	75.1	299	343
East Asia	816	1,175	1.46	0.96	67.6	74.8	1,475	1,712
South Asia	877	1,404	2.57	1.05	50.6	68.6	2,075	2,819
Europe	425	484	0.96	1.02	72.0	75.7	512	522
Oceania	16	23	1.39	1.10	65.6	73.8	30	36
USSR	214	265	1.16	1.10	69.6	74.6	310	355

SOURCE: United Nations (1981).

For most countries, net international migration is a relatively insignificant contributor to population change; consequently it is usually assumed to be zero for projection purposes. However, in a number of countries, for example, the United States, such a simplification is inadequate. Consequently, the United Nations' projections incorporate a contribution due to net international migration for 60 countries, where such migration was viewed to be nonnegligible in 1975-1980. This volume of net migration is progressively reduced to zero over time, except in those few countries where a continuation of current migration levels is likely for a considerable time in the future (for example, Mexico, Polynesia, and the United States).

Table 3.4 presents the United Nations' projections for major subregions and the world, as assessed in 1980.

In conclusion, a disaggregation of a national population by age is a fundamental dimension of the projection models used by international organizations such as the United Nations and the World Bank, and national agencies such as the U.S. Bureau of the Census. When applied to subnational populations, these models generally rely on projected age-specific net migration totals to link the demographic evolution of each regional population with that of the rest of the nation's population. This procedure, however, can introduce inconsistencies that will inflate or deflate national totals. A conceptually more elegant and practically more reliable method of projecting subnational populations is a "bottoms-up" approach that simultaneously focuses on all subpopulations and considers both their ages and their respective regions of residence. This is the topic of the next chapter.

4. MULTIREGIONAL POPULATION DYNAMICS: AGE AND LOCATION[1]

The evolution of an age-disaggregated multiregional population is governed by the history of fertility, mortality, and territorial mobility to which it has been subjected. This is true not only of the past but also of the future. Thus multiregional population projections are simply mathematical calculations that show the consequences, to a particular national population, of a set of assumptions regarding birth, death, and migration rates. Therefore, in the absence of an arithmetical error, the projections are correct irrespective of the future behavior of the population being projected. This is true because such projections are not predictions of what will actually happen, but are hypothetical exercises that arithmetically establish the numerical impacts of assumptions made regarding expected patterns of fertility, mortality, and migration. The consequences that the projections identify are conditional on the assumptions being fulfilled.

Demographers in the past have focused on uniregional populations that are assumed to be undisturbed by, or "closed to," migration. When migration was included at all, it usually was introduced by means of the notion of "net" migration. But net migrants are a statistical fiction, and net migration rates confound the impacts of changing migration propensities with those of changing population totals in origin and destination regions. Consequently, behavioral explanations of net migration may be misspecified, and spatial population analyses based on net migration rates are likely to mask important patterns underlying population redistribution.

These inadequacies of the uniregional approach led scholars in the mid-1960s to analyze the dynamics of spatial population change from a multiregional perspective. Subsequently, the marriage of multiregional projection models with multiregional life table models, and their expression in matrix form, to show transparently their natural correspondence with widely accepted conventional uniregional methods, established multiregional mathematical demography as a serious branch of analytical population studies (Rogers, 1975). The first two sections of this chapter set out the bare essentials of these projections and life table models. The models then are used to (1) examine once again the sources of urban growth; (2) simulate alternative patterns of population growth and urbanization in a prototype less developed country; and (3) project the future marital status and regional distributions of Sweden's female population.

The Multiregional Projection Model

To generalize the uniregional projection model (Chapter 3) to the case of several interacting regional populations (Chapter 2) one needs to replace scalars by matrices and vectors in all of the fundamental relationships defined in Chapter 3 (Rogers, 1975). Without loss of generality, we shall limit our exposition to the case of two regional populations, urban and rural, distinguishing them with one subscript in the case of population stocks and with two in the case of origin-destination-specific population flows. We begin by introducing a regional disaggregation in the uniregional projection model defined in equation 3.6. This may be achieved by replacing each scalar in that equation with either a matrix or a vector. Thus,

(1) each P(x;t) is replaced by the vector *P*(x;t), where

$$P(x;t) = \begin{bmatrix} P_u(x;t) \\ P_v(x;t) \end{bmatrix} \quad [4.1]$$

(2) each s(x) is replaced by the matrix *S*(x), where

$$S(x) = \begin{bmatrix} s_{uu}(x) & s_{vu}(x) \\ s_{uv}(x) & s_{vv}(x) \end{bmatrix} \quad \text{and} \quad [4.2]$$

(3) each b(x) is replaced by the matrix *B*(x), where

$$B(x) = \begin{bmatrix} b_{uu}(x) & b_{vu}(x) \\ b_{uv}(x) & b_{vv}(x) \end{bmatrix} \quad [4.3]$$

The single subscripts u and v refer, respectively, to urban and rural locations of stocks; a pair of subscripts denotes a flow from one location to another during the unit time interval. For example, $P_u(x;t)$ is the population aged x to x + 4 at last birthday residing in the urban region at time t; $s_{vu}(x)$ is the fraction living in the rural region when aged x to x + 4 at last birthday and in the urban region five years later; and $b_{uv}(x)$ is the average number of babies born during the unit age-time interval, and living in the rural region at the end of that interval, per person aged x to x + 4 at last birthday and living in the urban region at the start of the interval.

Figure 4.1 sets out a simple biregional, four-age-group numerical illustration. It can be viewed as a disaggregation of the numerical example presented in Chapter 3. Once again the reader should confirm that the matrix projection process ultimately stabilizes and exhibits both an unchanging growth ratio and a fixed age composition.

$$\begin{bmatrix} 198 \\ 198 \\ \hline 90 \\ 90 \\ \hline 90 \\ 90 \\ \hline 90 \\ 90 \end{bmatrix} = \left[\begin{array}{cc|cc|cc|cc} 0 & 0 & 1/2 & 1/4 & 2/3 & 1/6 & 1/8 & 1/8 \\ 0 & 0 & 1/4 & 1/2 & 1/6 & 2/3 & 1/8 & 1/8 \\ \hline 2/3 & 1/6 & 0 & 0 & 0 & 0 & 0 & 0 \\ 1/6 & 2/3 & 0 & 0 & 0 & 0 & 0 & 0 \\ \hline 0 & 0 & 2/3 & 1/6 & 0 & 0 & 0 & 0 \\ 0 & 0 & 1/6 & 2/3 & 0 & 0 & 0 & 0 \\ \hline 0 & 0 & 0 & 0 & 2/3 & 1/6 & 0 & 0 \\ 0 & 0 & 0 & 0 & 1/6 & 2/3 & 0 & 0 \end{array}\right] \begin{bmatrix} 108 \\ 108 \\ \hline 108 \\ 108 \\ \hline 108 \\ 108 \\ \hline 108 \\ 108 \end{bmatrix}$$

Figure 4.1 A Biregional Numerical Example

The generalized projection matrix G in equation 3.7 now takes on the form:

$$G = \begin{bmatrix} 0 & 0 & B(10) & B(15) & \dots \\ S(0) & 0 & 0 & 0 & \dots \\ 0 & S(5) & 0 & 0 & \dots \\ \cdot & \cdot & \cdot & & \cdot \\ \cdot & \cdot & & \cdot & \cdot \\ \cdot & \cdot & & & \cdot \end{bmatrix} \qquad [4.4]$$

and the fundamental matrix projection formula set out in 3.6 and in 2.15 remains valid, if

$$P(t) = \begin{bmatrix} P(0;t) \\ P(5;t) \\ \cdot \\ \cdot \\ \cdot \end{bmatrix} \qquad [4.5]$$

The matrices $S(x)$ and $B(x)$ are readily defined by replacing scalars with matrices in equations 3.16 and 3.29, respectively. Thus

$$S(x) = L(x+5)\,L(x)^{-1} \qquad [4.6]$$

and

$$B(x) = 1/2\,L(0)\,\ell(0)^{-1}\,[F(x) + F(x+5)\,S(x)] \qquad [4.7]$$

where

$$L(x) = \begin{bmatrix} {}_uL_u(x) & {}_vL_u(x) \\ {}_uL_v(x) & {}_vL_v(x) \end{bmatrix} \qquad \ell(0) = \begin{bmatrix} 100{,}000 & 0 \\ 0 & 100{,}000 \end{bmatrix}$$

$$F(x) = \begin{bmatrix} F_u(x) & 0 \\ 0 & F_v(x) \end{bmatrix}$$

and where the reciprocal notation of a minus one denotes matrix inversion (Rogers, 1971).

The matrix $L(x)$ is calculated by replacing scalars by matrices in equation 3.15 to obtain

$$L(x) = 5/2\,[\ell(x) + \ell(x+5)] \qquad [4.8]$$

where

$$\ell(x) = \begin{bmatrix} {}_u\ell_u(x) & {}_v\ell_u(x) \\ {}_u\ell_v(x) & {}_v\ell_v(x) \end{bmatrix}$$

and $L(x)$ is defined as before. The subscripts on the left side of $L(x)$ and $\ell(x)$ denote region of birth, those on the right side designate the region of current residence. Note that a second subscript on the right side is used to designate the region of future residence in variables such as $s(x)$, $b(x)$, and $p(x)$.

The successive $\ell(x)$ matrices may be computed by starting with 100,000 births in each region and surviving these two cohorts of babies forward, with probabilities of survival that are specific to region of residence at the start and end of the unit age interval:

$$\ell(x+5) = p(x)\,\ell(x) \qquad [4.9]$$

where the matrix of survival probabilities

$$p(x) = \begin{bmatrix} p_{uu}(x) & p_{vu}(x) \\ p_{uv}(x) & p_{vv}(x) \end{bmatrix}$$

should not be confused with the vector $P(t)$ that refers to the population stock at time t.

The element $p_{ij}(x)$ in the matrix $p(x)$ represents the probability that an individual in region i at exact age x will survive and be in region j five years later. Note that it does not denote the probability of making a move from i to j. Several moves may have been made during the unit age interval of five years. The life table is only interested in where members of the population were at the start and end of that interval.

Finally, we come to the generalization of the fundamental calculation that initiates the construction of a life table. Recalling equation 3.12, we replace scalars by matrices to obtain the matrix of survival probabilities (Rogers and Ledent, 1976):

$$p(x) = [I + 5/2\, M(x)]^{-1}\ [I - 5/2\, M(x)] \qquad [4.10]$$

where

$$M(x) = \begin{bmatrix} M_{ud}(x) + M_{uv}(x) & -M_{vu}(x) \\ -M_{uv}(x) & M_{vd}(x) + M_{vu}(x) \end{bmatrix}$$

and I is the identity matrix.

The element $M_{ij}(x)$ in matrix $M(x)$ denotes the age-specific migration rate from region i to region j; the element $M_{id}(x)$ denotes the age-specific death rate in region i. For the last open-ended age group, z years and over, the matrix $M(z)$ is used to close-out the life table:

$$L(z) = M(z)^{-1}\ \ell(z) \qquad [4.11]$$

Note that equation 4.11 is the multiregional generalization of equation 3.22.

Urbanization in India Revisited

The death rate in India's urban areas among 0- to 4-year olds in 1970 was, at 38.248 per thousand, considerably lower than the corresponding rate in rural areas, which stood at 57.712 per thousand (Appendix A). The outmigration rate for the same age group was 9.325 per thousand from urban areas and 5.552 per thousand from rural areas. Expressing these rates on a per capita basis, and collecting them to form the matrix $M(\mathrm{x})$ defined in equation 4.10, gives

$$M(0) = \begin{bmatrix} 0.038248 + 0.009325 & -0.005552 \\ -0.009325 & 0.057712 + 0.005552 \end{bmatrix}$$

and the associated matrix of transition probabilities

$$p(0) = [I + 5/2\,M(0)]^{-1}\;[I - 5/2\,M(0)] \qquad [4.12]$$

$$= \begin{bmatrix} 0.787865 & 0.021425 \\ 0.035989 & 0.727312 \end{bmatrix}$$

The probability matrix informs us that just over one out of fifty (0.021425) babies born in the rural areas will be living in urban areas 5 years later. Thus out of a rural birth cohort of 100,000 babies, 2,143 will be found residing in urban areas at exact age 5, another 72,731 will be living in rural areas, and 25,126 babies will have died before their fifth birthday. The corresponding totals for the urban birth cohort are 3,599 in rural areas, 78,786 in urban areas, and 17,615 dead. These results may be obtained by simple matrix multiplication, as defined in equation 4.9:

$$\ell(5) = p(0)\,\ell(0) \qquad [4.13]$$

$$= \begin{bmatrix} 0.787865 & 0.021425 \\ 0.035989 & 0.727312 \end{bmatrix} \begin{bmatrix} 100{,}000 & 0 \\ 0 & 100{,}000 \end{bmatrix}$$

$$= \begin{bmatrix} 78{,}786 & 2{,}143 \\ 3{,}599 & 72{,}731 \end{bmatrix}$$

Having found values for $\ell(0)$ and $\ell(5)$, we draw on equation 4.8 to calculate

$$L(0) = 5/2\,[\ell(0) + \ell(5)] \qquad [4.14]$$

$$= \begin{bmatrix} 446{,}966 & 5{,}356 \\ 8{,}997 & 431{,}828 \end{bmatrix}$$

An analogous set of calculations gives

$$L(5) = \begin{bmatrix} 384{,}001 & 13{,}857 \\ 23{,}792 & 355{,}099 \end{bmatrix}$$

whence, by equation 4.6,

$$S(0) = L(5)\,L(0)^{-1} \qquad [4.15]$$

$$= \begin{bmatrix} 0.858696 & 0.021439 \\ 0.036686 & 0.821862 \end{bmatrix}$$

The matrix $S(0)$ resembles $p(0)$, and so it should. Both represent similar transition probabilities; however, whereas the latter refers to populations at exact ages, the former refers to populations classified into age groups. Thus $p(0)$ is applied to $\ell(0)$ to yield $\ell(5)$ and $S(0)$ is applied to $L(0)$ to give $L(5)$. Each value in the matrix $S(0)$ should normally lie within the range defined by the two corresponding values in the matrices $p(0)$ and $p(5)$.

The four values calculated for the elements of $S(0)$ are set out in Table 4.1, along with corresponding values for all other age groups. They may be used to survive India's 1970 urban and rural populations (Appendix A). For example, recalling equations 4.4 and 4.5, we observe that

$$\begin{bmatrix} P_u(x+5;t+1) \\ P_v(x+5;t+1) \end{bmatrix} = \begin{bmatrix} s_{uu}(x) & s_{vu}(x) \\ s_{uv}(x) & s_{vv}(x) \end{bmatrix} \begin{bmatrix} P_u(x;t) \\ P_v(x;t) \end{bmatrix} \qquad [4.16]$$

TABLE 4.1 Fertility and Survivorship Elements of the Population Projection Matrix: India, 1970

	Fertility Elements				*Survivorship Elements*			
Age,x	$b_{uu}(x)$	$b_{uv}(x)$	$b_{vu}(x)$	$b_{vv}(x)$	$s_{uu}(x)$	$s_{uv}(x)$	$s_{vu}(x)$	$s_{vv}(x)$
0					0.858696	0.036686	0.021439	0.821862
5					0.957236	0.028591	0.018073	0.960723
10	0.069883	0.006145	0.003816	0.102052	0.946748	0.044240	0.034581	0.952089
15	0.252453	0.029194	0.018773	0.368544	0.904014	0.084133	0.071981	0.910900
20	0.378781	0.032549	0.021294	0.552139	0.900593	0.085333	0.071840	0.907690
25	0.360721	0.019368	0.012798	0.525156	0.936345	0.048957	0.036911	0.941430
30	0.283405	0.011788	0.008156	0.411861	0.947717	0.035121	0.025554	0.949044
35	0.174452	0.005915	0.004314	0.253223	0.950085	0.028518	0.020215	0.948075
40	0.083302	0.002677	0.001967	0.120781	0.945377	0.025801	0.017567	0.939788
45	0.026705	0.000538	0.000481	0.038792	0.935590	0.022110	0.014353	0.923247
50					0.912065	0.027142	0.016123	0.894760
55					0.869703	0.036763	0.020871	0.844249
60					0.810955	0.052643	0.028482	0.777584
65					1.959624*	0.366715	0.208395	1.449770*

SOURCE: Rogers (1982b).
*These elements exceed unity because they refer to survivorship into an open-ended age interval. Because not all members in that interval leave the population over a period of five years, a "correction" must be incorporated into the value of s(x).

which for x = 0 gives the following projected 1975 population aged 5 to 9 at last birthday:

$$\begin{bmatrix} 13{,}534{,}947 \\ 53{,}912{,}489 \end{bmatrix} = \begin{bmatrix} 0.858696 & 0.021439 \\ 0.036686 & 0.821862 \end{bmatrix} \begin{bmatrix} 14{,}140{,}200 \\ 64{,}966{,}800 \end{bmatrix}$$

Continuing in this manner for another 25 years, for all age groups except the first, gives rise to the projected population in the year 2000 that is set out in Table 4.2. The first age group is projected by applying the fertility submatrices $B(x)$ to the population in the childbearing age groups. These submatrices are defined by equation 4.7 and are readily calculated once the life table values for $L(0)$ and $S(x)$ are available. For example, the fertility submatrix associated with India's 20- to 24-year-old population is

$$B(20) = 1/2\, L(0)\, \ell(0)^{-1}\, [F(20) + F(25)S(20)] \qquad [4.17]$$

$$= \begin{bmatrix} 0.378781 & 0.021294 \\ 0.032549 & 0.552139 \end{bmatrix}$$

These four values of $b_{ij}(20)$ are presented in Table 4.1, together with corresponding values for all other childbearing age groups. Collectively, they define the fertility-survivorship behavior in India that produces the population in the first age group in the next time interval:

$$P(0;t+1) = B(10)P(10;t) + B(15)P(15;t) + \ldots + B(45)P(45;t) \qquad [4.18]$$

$$= \begin{bmatrix} 17{,}254{,}648 \\ 76{,}623{,}839 \end{bmatrix}$$

The results in Table 4.2 indicate that India's urban population in the year 2000 will total over 290 million persons, account for 27.7% of the national population total, and have been growing at the average rate of 2.74% per annum for the past five years (i.e., during the 1995-2000 time interval). Since the urban population growth rate will be declining over time, its value in the year 2000 will drop to 2.50% (Table 4.3A.1).

TABLE 4.2 Biregional Cohort-Survival Projection of the Urban and Rural Populations of India to the Year 2000

Age	*Population (in thousands)**			*Age Composition**		
x	*Total*	*Urban*	*Rural*	*Total*	*Urban*	*Rural*
(1)	*(2)*	*(3)*	*(4)*	*(5)*	*(6)*	*(7)*
0	163,181	37,231	125,950	0.1553	0.1280	0.1657
5	127,799	31,167	96,632	0.1216	0.1072	0.1272
10	115,345	28,445	86,900	0.1098	0.0978	0.1144
15	102,424	26,237	76,187	0.0975	0.0902	0.1003
20	87,095	24,619	62,474	0.0829	0.0847	0.0822
25	74,851	22,608	52,242	0.0712	0.0777	0.0687
30	61,813	19,281	42,532	0.0588	0.0663	0.0560
35	74,163	22,884	51,279	0.0706	0.0787	0.0675
40	60,587	19,488	41,100	0.0577	0.0670	0.0541
45	41,023	13,735	27,288	0.0390	0.0472	0.0359
50	35,698	11,616	24,082	0.0340	0.0399	0.0317
55	31,563	9,182	22,381	0.0300	0.0316	0.0295
60	24,962	7,066	17,896	0.0238	0.0243	0.0236
65	19,072	5,566	13,507	0.0182	0.0191	0.0178
70+	31,118	11,676	19,443	0.0296	0.0402	0.0256
Total	1,050,692	290,799	759,892	1.0000	1.0000	1.0000
Share	1.0000	0.2768	0.7232			
Annual growth rate	0.0210	0.0274	0.0186			

*Slight differences are due to independent rounding.

The Sources of Urban Growth Once Again

In Chapter 1 we examined the demographic sources of urban growth using the uniregional model without a disaggregation by age. In Chapter 2 we generalized the uniregional model into a biregional one and looked at the question once again. There we concluded that the simple biregional model's guarantee of an ultimately declining relative contribution of migration seriously limited its usefulness for resolving the question of whether it is natural increase or net migration that is the principal source of urban population growth during the urbanization transition. A more realistic model was called for, and we now propose to look at the matter once more using an age-disaggregated biregional model.

Table 4.3 sets out age-specific population projections for India and for the Soviet Union. The projections show that exposing India to the migration rates of the Soviet Union would urbanize India in 50 years to the level ultimately attained by the Soviet Union, whereas introducing India's migration rates into the Soviet Union's growth regime would rapidly "deurbanize" that national population.

TABLE 4.3 Age-Disaggregated Projections of Observed Populations Exposed to Different Regimes of Growth: India and the Soviet Union

A. India's Population								
1. India's Growth Regime					*2. India's Natural Increase Rates with Soviet Union's Migration Rates*			
100U	r_u	m_u	$m_u/r_u \times 100$	*T*	*100U*	r_u	m_u	$m_u/r_u \times 100$
19.9	0.037	0.017	47.1	1970	19.9	0.175	0.155	88.8
21.6	0.035	0.015	42.8	1975	33.4	0.087	0.058	67.1
23.3	0.033	0.014	41.2	1980	44.4	0.065	0.038	57.9
27.7	0.025	0.009	33.8	2000	69.8	0.026	0.007	26.2
30.1	0.023	0.006	28.2	2020	77.1	0.017	0.002	10.9
33.8	0.019	0.004	19.8	Stability	79.0	0.014	0.001	3.8
B. Soviet Union's Population								
1. Soviet Union's Growth Regime					*2. Soviet Union's Natural Increase Rates with India's Migration Rates*			
100U	r_u	m_u	$m_u/r_u \times 100$	*T*	*100U*	r_u	m_u	$m_u/r_u \times 100$
56.3	0.025	0.016	63.8	1970	56.3	0.004	–0.005	–132.8
60.5	0.020	0.012	59.5	1975	54.5	0.003	–0.004	–156.9
64.4	0.018	0.011	57.4	1980	52.8	0.002	–0.003	–167.4
73.4	0.005	0.003	60.9	2000	45.6	–0.002	–0.002	87.0
76.9	0.004	0.002	44.9	2020	39.6	0.002	0.000	27.3
77.5	0.002	0.001	72.7	Stability	29.3	0.010	0.007	71.8

SOURCE: A. Rogers (1982a). "Sources of Urban Population Growth and Urbanization, 1950-2000: A Demographic Accounting." *Economic Development and Cultural Change* 30 (3): 483-506. Copyright: University of Chicago Press, 1982. Reprinted with permission.

The introduction of age composition alters the results in favor of migration as a contributor to urban growth. In the Indian case it increases migration's ultimate contribution threefold (from 6.0% to 19.8%); in the Soviet Union example it reverses the ranking itself, making migration the principal source of urban growth. What accounts for this reversal?

The disaggregation by age does not change the pattern of evolution of the aggregate urban net inmigration rate $m_u(t)$. In both the Indian and the Soviet illustrations it declines sharply from its initial level. But now the aggregate rate of natural increase no longer remains constant, dropping from 2% to 1.5% in the case of India and from 0.9% to 0.05% in the case of the Soviet Union. The cause of this decline in the aggregate rate is, of course, the gradual aging of the population and the associated shift in its age composition. This shift alters the relative weights with which the fixed age-specific rates are consolidated to form the aggregate crude rates. The net result is an increased relative contribution of net migration as a source of urban population growth, a consequence apparently of the fact that, as with mortality (but not with fertility), the risks of migration are experienced by individuals of all ages.

Table 4.3 illustrates the short-run impacts of high rates of rural to urban migration on urban natural increase. In Table 4.3A.2, for example, the crude rate of urban natural increase, $r_u - m_u$, fixed at 20 per 1,000 in Table 2.1, now approaches 29 per 1,000 in 1975 (0.087 – 0.058 = 0.029) and 27 per 1,000 in 1980 (0.065 – 0.038 = 0.027) before declining to roughly half those levels in the subsequent decades. Nevertheless, the even higher short-run rates of net urban inmigration ensure the primacy of migration as a source of urban growth for over two decades. Observe that increasing rural to urban migration still produces an ultimately lower urban growth rate, but now only after migration ceases to be the principal source of urban population growth—a crossover that, in the illustration, occurs when the national population is about 50% urban.

In conclusion, it appears that the principal effect of introducing age composition into the fixed-rate projection model is to decrease the aggregate rate of natural increase over time, while slowing down the decline of the urban net migration rate. Because these two contributors to urban growth now can exhibit different rates of decline over time, their relative importance as sources of urban growth also can change, and in patterns that are difficult to anticipate.

The decompositions presented here have attempted to identify the instantaneous contributions of migration and natural increase to urban population growth over time. The focus has been on estimating the fraction of growth at each moment, t, that could be attributed to the

TABLE 4.4 Age-Disaggregated Place-of-Residence-by-Place-of-Birth Projections of Observed Populations Exposed to Different Regimes of Growth: India and the Soviet Union

A. India's Population

1. India's Growth Regime				*2. India's Natural Increase Rates with Soviet Union's Migration Rates*		
$100U$	$100U_N$	$100U_A$	T	$100U$	$100U_N$	$100U_A$
19.9	19.9	0	1970	19.9	19.9	0
21.6	19.1	2.6	1975	33.4	20.7	12.8
23.3	18.6	4.6	1980	44.4	23.8	20.6
27.7	18.6	9.1	2000	69.8	39.5	30.3
30.1	19.4	10.7	2020	77.1	51.3	25.7
33.8	23.6	10.2	Stability	79.0	66.9	12.1

B. Soviet Union's Population

1. Soviet Union's Growth Regime				*2. Soviet Union's Natural Increase Rates with India's Migration Rates*		
$100U$	$100U_N$	$100U_A$	T	$100U$	$100U_N$	$100U_A$
56.3	56.3	0	1970	56.3	56.3	0
60.5	53.9	6.6	1975	54.5	52.9	1.6
64.4	53.2	11.3	1980	52.8	50.0	2.8
73.4	53.5	20.0	2000	45.6	39.5	6.2
76.9	55.2	21.8	2020	39.6	31.0	8.6
77.5	61.2	16.4	Stability	29.3	16.5	12.8

SOURCE: A. Rogers (1982a). "Sources of Urban Population Growth and Urbanization, 1950-2000: A Demographic Accounting." *Economic Development and Cultural Change* 30 (3): 483-506. Copyright University of Chicago Press, 1982. Reprinted with permission.

migration or natural increase rates prevailing at that same moment. But migrants bear children, and it may be desirable to identify that particular contribution to urban growth more explicitly in efforts to answer the question of whether it is migration or natural increase that is the major source of urban population growth. A convenient way of approximating this contribution is to disaggregate the projection model further to permit it to keep track of the respective places of birth of the projected populations.

A number of studies dealing with the urban problems of the less developed world today view with concern the high fractions of urban residents born in rural areas, implying that these high fractions of "lifetime migrants" reflect high rates of rural-to-urban migration. Table 4.4 indicates that this may not necessarily be true by presenting the results of a further disaggregation of the age-specific projections summarized earlier in Table 4.3. The additional disaggregation is by place of birth (Philipov and Rogers, 1981). Because no data are available to disaggregate the initial 1970 population along this dimension, we focus only on the allocation that evolves at stability, inasmuch as this result is independent of the starting condition and is a function only of the particular growth regime.

The place-of-residence-by-place-of-birth (PRPB) projections demonstrate that the existence of a large fraction of rural-born urban residents is not necessarily an indication of high rural-to-urban migration rates. Indeed, the association is apparently the other way around. High rates of rural-urban migration, such as those experienced in the Soviet Union, for example, generate urban populations with a higher share of urban-born "natives" than do lower migration rates, such as those found in India. The reason for this apparent paradox is, once again, the influence of the urbanization level, U(t).

High rates of net urban inmigration produce high levels of urbanization, with the result that urban areas account for increasingly larger fractions of national births over time. For example, on 1970 rates, roughly three-fourths of all national births in the Soviet Union occur in urban areas at stability, compared with only one-fourth in India. This situation gives rise to a high fraction of natives in urban areas and explains why only 21% of the Soviet Union's stable urban population is rural born, compared with India's 30%.

The place-of-birth disaggregation can be carried one step further by keeping track of the place of birth of the parent as well as that of the child. Such a projection disaggregates the native urban population into two parts, separating the first-generation natives (urban-born children of rural-born parents) from the rest.

On 1970 rates, a projection that separately identifies first-generation natives shows that of the 23.6% urban natives in India at stability, over a third (36.0%) are children of rural-born parents; whereas of the 61.2% urban natives in the Soviet Union at stability, only about a fifth (21.9%) fall into this category. Thus, if one includes the children of rural lifetime migrants into the accounting, more than half (55.3%) of India's ultimate urban population will consist of lifetime migrants and their direct (first-generation) contribution to urban natural increase. The corresponding result for the Soviet Union is only 38.5%.

Our major conclusion regarding the sources of urban growth is that this fundamental question does not have a simple unequivocal answer. At different periods during a nation's urbanization transition, its urban population may grow primarily as a consequence of net urban inmigration; at other times the main contributor may be urban natural increase. The "guaranteed" ultimate decline of the relative contribution of migration projected by the simple model that disregards age is merely a direct consequence of the model specification, which ignores the effects of age distribution.

The analysis of the demographics of urbanization and the changing contributions of natural increase and migration over time lead us to put forward a few important observations:

(1) The principal effect of migration is to establish the level of urbanization, whereas that of natural increase is to determine the rate of urban population growth.
(2) Although a sharp increase in the rate of rural to urban migration temporarily raises the urban population growth rate, its ultimate effect is to urbanize the population more rapidly and thereby depress the urban growth rate to a lower level than it would have reached in the absence of the increase.
(3) The relative importance of the two sources of urban population growth and urbanization may differ depending on whether the focus is on periodical net additions to the urban population stock or on the changing projected composition of that stock, for example, the disaggregation between natives and lifetime migrants.
(4) A relatively large fraction of rural-born people among urban residents is not necessarily a sign of high rural to urban migration rates.

Scholars and policymakers often disagree when it comes to evaluating the desirability of current rates of rapid urban population growth and rural-urban migration in the less developed world. Some see these trends as effectively speeding up national processes of socioeconomic development, whereas others believe their consequences to be largely undesirable and argue that both trends should be slowed down.

Regardless of how desirable or feasible it may be to restrict the movement of people in the interests of national welfare, it seems reasonable to ask whether such efforts could have a significant impact on the growth rates of urban centers. Our simple decompositions do not provide a clear-cut answer, but they nevertheless do cast some doubt on the matter, inasmuch as they indicate that slowing down rural-to-urban migration is not likely to produce more than a short-run reduction of urban population growth rates unless fertility levels are also reduced.

Simulating the Urbanization Transition

Urbanization is a finite process all nations go through in their transition from an agrarian to an industrial society. Such urbanization transitions can be depicted by attenuated S-shaped curves (e.g., logistic curves) that tend to show a swift rise in the proportion urban around 20%, a flattening out at a point somewhere between 40% and 60%, and a halt or even a decline in this proportion at levels above 75%. They are characterized by distinct urban-rural differentials in fertility-mortality levels and patterns of decline, and by a massive net transfer of population from rural to urban areas through internal migration.

In a now classic analysis of the demoeconomic consequences of fertility reduction, Ansley Coale (1969) examined some of the ways in which the population characteristics of less developed countries are related to their poverty and how alternative demographic trends might affect their development. Coale was concerned with the implication of alternative possible future courses of fertility for the growth in per capita income and for the provision of productive employment. He examined two specific alternatives: the maintenance of fertility at its current level and a rapid reduction in fertility, over a transitional period of about 25 years, amounting to 50% of the initial level.

After generating the two alternative projections or "scenarios," Coale went on to consider the effects that these two different trends in fertility would have on three important population characteristics: the burden of dependency, defined as the total number of persons in the population divided by the number of persons aged 15 to 64; the annual rate of increase of the labor force, defined as the growth rate of the population aged 15 to 64; and population density, defined as the number of persons in the labor force ages relative to land area and other resources.

A recent generalization of Coale's scenario-building approach focuses on some of the demoeconomic consequences of rapid urbanization (Rogers, 1978b). It begins by developing four alternative population scenarios and then goes on to examine the implications that alternative trends in migration and fertility would have on Coale's three important population characteristics: the dependency burden, the growth rate of labor force "eligibles," and the density of the population.

As in the Coale paper, a hypothetical initial population of one million persons with an age composition and fertility-mortality rates typical of a Latin American country was projected 150 years into the future. To his alternative projections (A. fertility unchanged and B. fertility reduced), however, two others were added by varying the assumptions on internal migration (a. migration unchanged and b. migration increased). This produced the following four possible combinations:

	a. Migration unchanged	b. Migration increased
A. Fertility unchanged	Projection Aa	Projection Ab
B. Fertility reduced	Projection Ba	Projection Bb

Coale's assumptions about initial and future patterns of mortality and fertility were a crude birth rate of about 44 per 1,000 and a crude death rate of 14 per 1,000, giving rise to a population growing at 3% per year. Starting with an expectation of life at birth of approximately 53 years, he assumed that during the next 30 years it will rise to about 70 years, at which point no further improvement will occur. In Coale's Projection A current age-specific rates of childbearing are fixed for 150 years; in Projection B they are reduced by 2% each year for 25 years (reducing fertility to half of its initial level), at which point they too are fixed for the remainder of the projection period.

In the four urbanization scenarios, Coale's data and assumptions were spatially disaggregated in the following manner. Twenty percent of the initial population of a million persons was taken to be urban. The initial values for birth and death rates were assumed to be lower in urban areas than in rural areas (40 against 45 per 1,000 for the birth rate, and 11 against 15 per 1,000 for the death rate). Mortality and fertility were reduced as in the Coale projections, but the declines were assumed to occur ten years sooner in urban areas (25 instead of 35 years for the decline in mortality, and 20 instead of 30 years for the decline in fertility).

Initial rates of outmigration were set equal to those prevailing in India in 1970 (Chapter 2); that is, a crude outmigration rate from urban areas of 10 per 1,000 and a corresponding rate from rural areas of 7 per 1,000. The age-specific rates of outmigration from urban areas were held fixed in all four projections, as were the corresponding rates from rural areas in the two "a" projections. Outmigration from rural areas in the two "b" projections, however, was assumed to increase six-fold over a period of 50 years and then to drop to half its peak value over the following 30 years, after which it was held unchanged for the remaining 70 years of the projection period.

The assumptions appear to be reasonable in that the hypothetical urbanization paths they chart are plausible. For example, the percentage-urban paths for the "b" projections resemble the general shape of historically observed urbanization paths, and the trajectories of urban and rural growth rates for these projections are in general similar to those exhibited by data for several developed nations.

As in Coale's scenarios, the initial population and the future regime of mortality are the same for all of the four population projections summarized in Figure 4.2. The major impact of the drop in fertility appears in the projected totals: the "A" projection totals are about 24 times as large as the "B" projection totals after 150 years. Migration's impact, on the other hand, appears principally in the spatial distribution of these totals: the "a" projections allocate approximately a third of the national population to urban areas after 150 years, whereas the "b" projections double this share.

The principal demographic impacts of reduced fertility described by Coale are not altered substantially by the introduction of migration as a component of change and by the concomitant spatial subdivision of the national population into urban and rural sectors. Figures 4.3 and 4.4 show that for a given regime of migration (a or b), the major impacts of reduced fertility are, as in the Coale model, a decline in the burden of dependency in the short run, a lowering of the growth rate of the labor force population in the medium run, and a very much lower density of people to resources in the long run. The spatial model, however, does bring into sharp focus urban-rural differentials in: (1) dependency burdens and the relative magnitudes of their decline following fertility reduction and (2) initial growth rates of the labor force populations and the paths of their gradual convergence in the long run.

The dependency ratio in urban areas is 19 points lower than its rural counterpart at the start of the projection period. With constant fertility, the regional dependency burdens remain essentially unchanged. Declining fertility, however, narrows these differentials to almost a third of their original values, as the urban drop of 33 points is matched by a corresponding decline of 45 points in rural areas.

The annual growth rates of the labor force populations in urban and rural areas initially are 0.05 and 0.03, respectively. For both migration regimes, however, they converge to approximately the same values in the long run: 0.04 in the constant fertility scenarios and slightly above 0.01 in the reduced fertility projections.

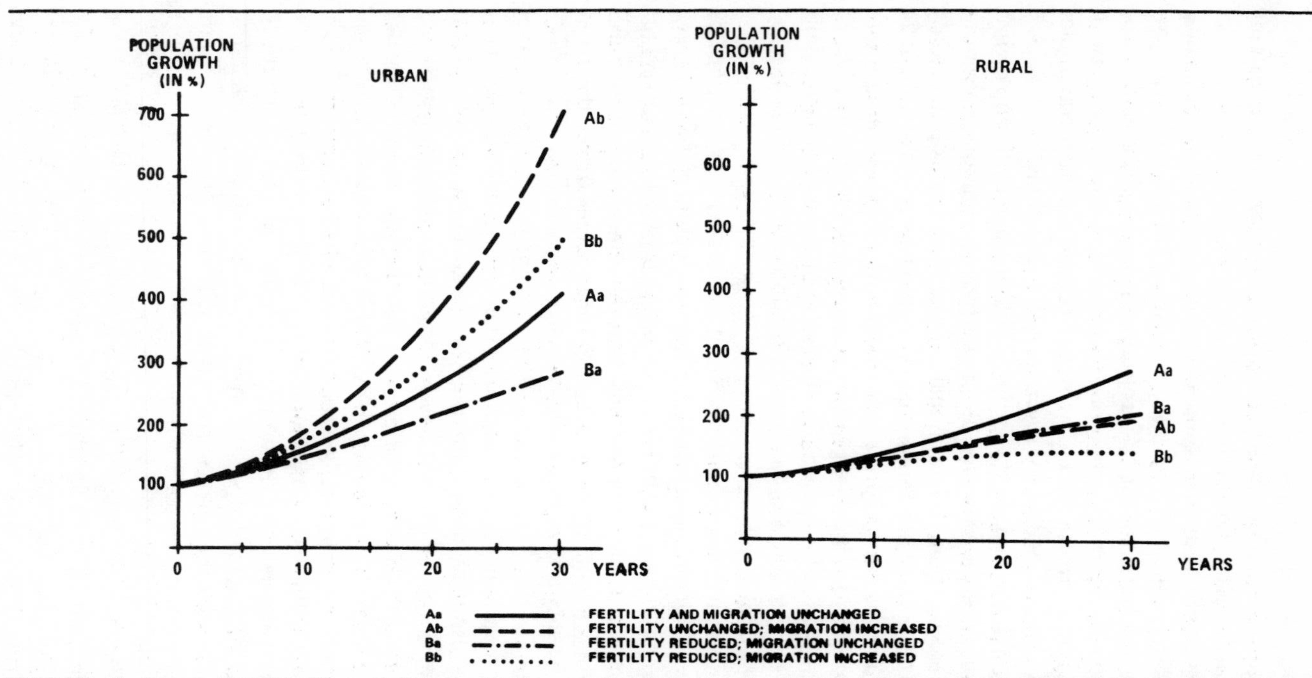

SOURCE: A. Rogers (1978b). From McMains and Wilcox's **Alternatives for Growth: The Engineering and Economics of Natural Resources Development**, Copyright 1978 by the National Bureau of Economic Research, Inc. Reprinted by permission of Ballinger Publishing Company.

Figure 4.2 Alternative Projections of the Population of a Less Developed Country: Four Scenarios

The major demographic impacts of increased rural-urban migration for a given regime of fertility, as set out in Figure 4.3 and 4.4, are negligible with respect to dependency burdens and are of paramount importance, in the short and medium runs, with regard to the growth rate of the population aged 15 to 64. In the long run, migration also has a moderately powerful impact on the density of workers to resources in rural areas.

Perhaps the most interesting observation suggested by the scenarios is the transitory nature of high rates of urban growth. In the "b" projections, urban growth rates in excess of 6% per annum occur only in the short run, as the national population is in its early phases of urbanization. This sudden spurt of growth of urban areas in the short run declines over the medium run, and in the long run levels off at a rate below that generated by the fixed migration regime. The growth curve of rural areas, of course, assumes a reverse trajectory, with the growth of the rural working population declining to relatively low, even negative, levels before increasing to stabilize at about the same level as that prevailing in the urban population.

Increased migration into cities reduces the size of rural populations and hence their density with respect to rural resources such as agricultural land. Figures 4.3 and 4.4 show that the relative size of the rural population aged 15 to 64 is over 2.5 times larger under the fixed migration schedules of projections "a" than under the increased rural-urban migration rates of projections "b." Thus the "b" scenarios create rapid urban growth and exacerbate human settlement problems, but at the same time they reduce the density of rural populations to land and other rural resources. The "a" scenarios, on the other hand, give urban areas more time to cope with growth, but they do so at the cost of increasing rural population densities. "Hyperurbanization" and "rural overpopulation," therefore, are the two sides of a fundamental policy question regarding spatial development.

Multistate Projections of Sweden's Female Population

The mathematical apparatus for tracing the demographic consequences of movements of people between urban and rural regions is the same as that for assessing the impacts of their movements between different states of existence: for example, married to nonmarried, employed to unemployed, healthy to sick. This recognition has had a profound impact on formal demography (Rogers, 1980; Land and Rogers, 1982). It has produced a powerful generalization of conven-

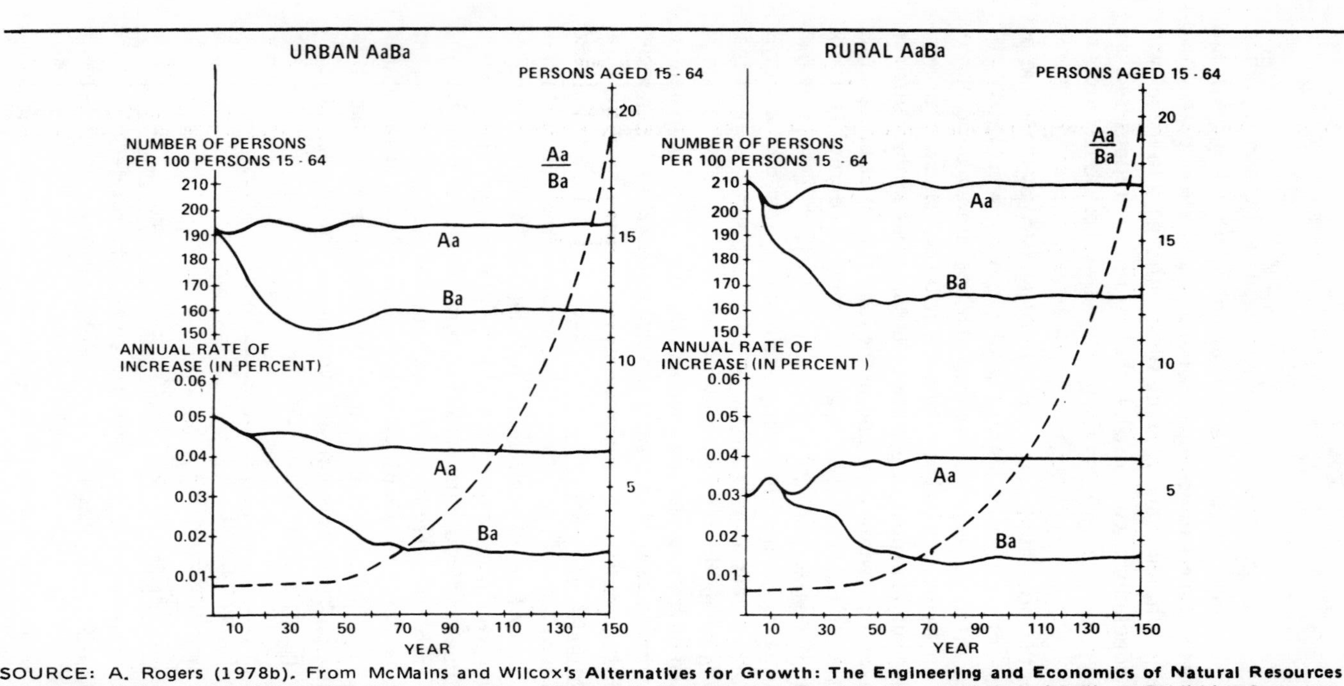

SOURCE: A. Rogers (1978b). From McMains and Wilcox's **Alternatives for Growth: The Engineering and Economics of Natural Resources Development**, Copyright 1978 by the National Bureau of Economic Research, Inc. Reprinted by permission of Ballinger Publishing Company.

Figure 4.3 Dependency Burden, Annual Rate of Increase, and Relative Size of Population Aged 15-64 Years: Alternative Urban-Rural Projections, Migration Unchanged

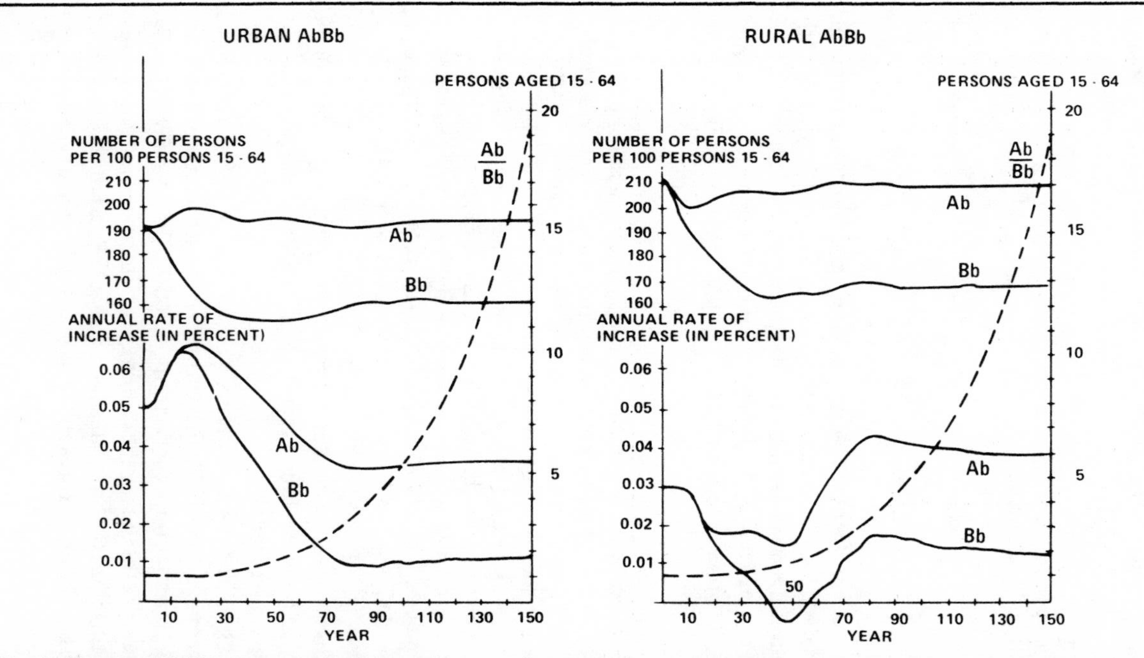

SOURCE: A. Rogers (1978b). From McMains and Wilcox's **Alternatives for Growth: The Engineering and Economics of Natural Resources Development**, Copyright 1978 by the National Bureau of Economic Research, Inc. Reprinted by permission of Ballinger Publishing Company.

Figure 4.4 Dependency Burden: Annual Rate of Increase, and Relative Size of Population 15-64 Years: Alternative Urban-Rural Projections, Migration Increased

TABLE 4.5 Patterns of Marital Status Change, Females, Sweden, 1974

From/To	*Single*	*Married*	*Widowed*	*Divorced*	*Deaths*	*Births*	*Population*
Single	–	37,768	–	–	7,562	15,257	1,659,430
Married	–	–	24,961	27,966	11,847	36,519	1,890,436
Widowed	–	399	–	–	17,797	119	385,070
Divorced	–	6,342	–	–	1,548	1,305	163,599
Total					38,754	53,200	4,098,535

SOURCE: Appendix B.

tional demographic techniques for studying the transitions that people experience over their lifetime as they progress from birth to death. Governmental agencies such as the U.S. Bureau of Labor Statistics (1982) have adopted this methodology, and the new *International Encyclopedia of Population* describes it as a fundamental new departure that brings

> many demographic analyses under a single approach. In this approach, individuals are permitted to move freely within a matrix, among several states or conditions. Transitions can occur from any state to any other state and in either direction. A ready example is the geographic one, where movement can literally be among all "states," but statistically speaking, movement can be as readily from employment to unemployment or from marriage to nonmarriage. The highly sophisticated mathematical equipment already developed in the matrix field greatly expands the potential for analysis of these events [Ross, 1982, p. 424].

This section considers an illustration in which the "migration" of the population is between four "regions" of existence: single, married, widowed, and divorced.

According to Table 4.5, the female population of Sweden increased by 14,446 people during 1974. Starting the year with a total of 4,098,535 women, the population experienced 53,200 births of baby girls and 38,754 female deaths during the ensuing year (international migration is ignored in this illustration). Thus the total at the end of the year stood at 4,112,981 persons. Expressed in crude birth and death rates, we have that

$$\begin{aligned} P(1975) &= (1 + b - d)\, P(1974) \\ &= (1 + r)\, P(1974) \\ &= (1 + 0.003525)\, 4{,}098{,}535 \\ &= 4{,}112{,}981. \end{aligned} \qquad [4.19]$$

During the year, 97,436 women, 2.38% of the female population, changed their marital status (Table 4.5) with marriages accounting for 45.68% of the changes, widowhoods for 25.62%, and divorces for 28.70%. First marriages amounted to 84.85% of the total number of marriages.

The data in Table 4.5 may be expressed in the form of the matrix projection model defined in Chapter 2. Instead of migrations between regions, we have movements between states. The accounting equations now assert that the population in each marital state at the end of the year is equal to the population at the start of the year, minus deaths and movements out of the state, plus movements into the state. In the case of the single (never-married) population, $P_s(t)$ say, there is also the increment due to births. For example,

$$\begin{aligned} P_s(1975) &= (1{,}659{,}430 - 7{,}562 - 37{,}768 + 15{,}257) \\ &\qquad + 36{,}519 + 119 + 1{,}305 \\ &= 1{,}667{,}300 \end{aligned}$$

which expressed in rates is

$$\begin{aligned} P_s(1975) &= (1 - 0.004557 - 0.022760 + 0.009194)\ 1{,}659{,}430 \\ &\quad + 0.019318\ (1{,}890{,}436) + 0.000309\ (385{,}070) \\ &\quad + 0.007977\ (163{,}599) \\ &= 1{,}667{,}300 \end{aligned}$$

Collecting the four subpopulations into a vector $P(t)$, we may define the familiar matrix projection model:

$$P(t+1) = G\,P(t) \qquad [4.20]$$

or

$$\begin{bmatrix} 1{,}667{,}300 \\ 1{,}870{,}171 \\ 391{,}835 \\ 183{,}675 \end{bmatrix} = \begin{bmatrix} 0.981878 & 0.019318 & 0.000309 & 0.007977 \\ 0.022760 & 0.965736 & 0.001036 & 0.038766 \\ 0 & 0.013204 & 0.952746 & 0 \\ 0 & 0.014793 & 0 & 0.951772 \end{bmatrix} \begin{bmatrix} 1{,}659{,}430 \\ 1{,}890{,}436 \\ 385{,}070 \\ 163{,}599 \end{bmatrix}$$

Classifying the Swedish female population by marital status is a useful form of disaggregation because it illuminates patterns of marital status change. Classifying the same population by residential status identifies patterns of spatial redistribution. For example, in 1974 766,565 of the 4,098,535 Swedish women lived in Stockholm, the capital city (Appendix B). Among these 14,726 migrated to the rest of Sweden during the year and 12,858 migrated in the reverse direction. The Stockholm population experienced 6,640 deaths and 9,991 births; the corresponding totals for the rest of Sweden were 32,114 and 43,209, respectively. Thus the following biregional projection model describes population redistribution during that period:

$$\begin{bmatrix} 768{,}048 \\ 3{,}344{,}933 \end{bmatrix} = \begin{bmatrix} 0.985161 & 0.003859 \\ 0.019210 & 0.999471 \end{bmatrix} \begin{bmatrix} 766{,}565 \\ 3{,}331{,}970 \end{bmatrix}$$

Combining the classification by marital status with that of location gives rise to eight states and an 8 by 8 projection matrix. The reader should use the data in Appendix B to create such a multistate model.

All of the preceding has ignored age. Incorporation of that added dimension into the analysis is straightforward and follows the procedures developed earlier in this chapter. Tables 4.6 and 4.7 set out the resulting projections to the year 2000. Note the aggregation bias that is introduced by the alternative consolidations of the full eight-state age-disaggregated model.

Multiregional/Multistate Demography: A New Perspective

A number of pressing national and regional population issues arise as a consequence of unanticipated patterns of change in the age composition, spatial distribution, and group status of population stocks. These changes generally evolve slowly, but their effects are widely felt, and the problems they bring in their wake typically are lasting and complex. Public awareness and public action are slow in coming, and all too often both are stimulated by an inadequate understanding of the processes generating the patterns of change.

Demographers have addressed these issues and have sought to understand the associated underlying processes, but their analytical apparatus has been inadequate. A particular shortcoming of this apparatus has been its central focus on the evolution of a single population as it develops over time, while being exposed to sex- and age-specific rates of events, such as births and deaths (Long, 1981). Such a unistate perspective of population growth and change is ill-equipped to examine

TABLE 4.6 Multistate Population Projections (in thousands): Sweden, Females, 1974-2004-Stability

	Population Projections								
	1–State	*2–States*			*4–States*				
Variable	*Sweden Total*	*Stockholm*	*R. Sweden*	*Sweden Total*	*Never Married*	*Married*	*Widowed*	*Divorced*	*Sweden Total*
1974									
Population	4,099	767	3,332	4,099	1,659	1,890	385	164	4,099
Mean age	38.6	38.2	38.7	38.6	20.5	47.0	70.5	49.7	38.6
Share	100.0	18.7	81.3	100.0	40.5	46.1	9.4	4.0	100.0
2004									
Population	4,120	740	3,381	4,121	1,659	1,528	417	434	4,038
Mean age	41.1	41.5	41.0	41.1	23.8	48.5	73.8	54.8	41.6
Share	100.0	18.0	82.0	100.0	41.1	37.8	10.3	10.7	100.0
Growth rate	–0.0014	–0.0028	–0.0011	–0.0014	0.0001	–0.0067	–0.0085	0.0088	–0.0024
Stable									
Mean age	41.8	43.0	42.8	42.8	30.3	48.6	73.7	58.7	44.2
Share	100.0	17.1	82.9	100.0	43.7	34.9	9.5	11.9	100.0
Growth rate	–0.0016	–	–0.0041	–	–	–	–0.0064	–	–

SOURCE: Rogers and Planck (1984).

TABLE 4.7 Multistate Population Projections (in thousands): Sweden, Females, 1974-2004-Stability

	8-State Population Projection								
	Stockholm				*Rest of Sweden*				
Variable	*Never Married*	*Married*	*Widowed*	*Divorced*	*Never Married*	*Married*	*Widowed*	*Divorced*	*Sweden Total*
1974									
Population	304	364	68	31	1,355	1,527	317	132	4,099
Mean age	20.6	46.0	70.2	48.8	20.5	47.2	70.6	49.9	38.6
Share	7.4	8.9	1.7	0.8	33.1	37.3	7.7	3.2	100.0
2004									
Population	282	260	72	114	1,379	1,274	346	313	4,041
Mean age	24.2	47.4	74.1	53.5	23.7	48.7	73.7	55.4	41.6
Share	7.0	6.4	1.8	2.8	34.1	31.5	8.6	7.8	100.0
Growth rate	–0.0016	–0.0087	–0.0125	0.0086	0.0007	–0.0061	–0.0075	0.0086	–0.0023
Stable									
Mean age	29.9	47.2	73.5	57.6	30.2	48.9	73.7	59.1	44.1
Share	7.0	5.7	1.5	3.0	36.7	29.4	8.1	8.6	100.0
Growth rate	–	–	–	–	–0.0063	–	–	–	–

SOURCE: Rogers and Planck (1984).

the evolution of a system of interacting populations that are linked by gross flows between various states of existence.

This last chapter has outlined a multiregional/multistate model of population growth and change. In addition to events such as births and deaths, this perspective focuses on gross flows and on multiple interacting populations. It uses these as numerators and denominators, respectively, to define rates of occurrence that refer to populations exposed to the possibility of experiencing such occurrences, that is, occurrence/exposure rates.

Two important consequences follow. First, the multiregional/multistate approach avoids potential inconsistencies arising from inappropriately defined rates. Second, it allows one to follow individuals across several changes of states of existence, thereby permitting the disaggregation of current or future population stocks and flows by previous states of existence.

A focus on occurrences of events and transfers, and their association with the populations that are exposed to the risk of experiencing them, enhances our understanding of, for example, patterns of fertility, mortality, and migration. By not permitting such an association, the unistate approach can produce undesirable biases.

Heterogeneous populations contain subgroups for which demographic behavior is diverse. To the extent that these diverse behaviors can be incorporated in a formal analysis, illumination of the aggregate patterns is enhanced. For instance, our understanding of marital dissolution is enriched by information on the degree to which divorces occur among those previously divorced. In generating such information, a multiregional/multistate analysis can identify, for example, how much of the current increase in levels of divorce in many countries can be attributed to "repeaters" as opposed to "first-timers."

Multiregional/multistate demography is a young branch of formal demography, and its potential contributions are only now coming to be recognized. Further progress in the field will depend on the availability of the necessary disaggregated data for carrying out the analyses and projections that would promote its further development and acceptance.

NOTE

1. Portions of this chapter were reprinted from my chapter in McMains and Wilcox's *Alternatives for Growth: The Engineering and Economics of Natural Resources Development*, Copyright 1978 by the National Bureau of Economic Research, Inc. Reprinted by permission from Ballinger Publishing Company.

APPENDIX A: Demographic Data for India

URBAN REGION

Age	Population	Births	Deaths	Outmig.	Observed rates (per 1000) Birth	Death	Outmig.
0	14,140,200		540,830	131,860		38.248	9.325
5	14,798,300		58,278	98,442		3.938	6.652
10	13,637,500		23,598	70,661		1.730	5.181
15	10,944,900	361,195	20,245	151,924	33.001	1.850	13.881
20	10,454,900	923,207	29,320	253,459	88.304	2.804	24.243
25	8,955,700	805,956	24,581	109,872	89.994	2.745	12.268
30	7,612,400	580,051	23,620	63,759	76.198	3.103	8.376
35	6,881,500	367,275	25,868	43,812	53.371	3.759	6.367
40	5,714,300	148,412	27,618	32,235	25.972	4.833	5.641
45	4,476,500	53,492	30,450	23,847	11.950	6.802	5.327
50	3,810,300		39,787	15,975		10.442	4.193
55	2,223,400		32,371	18,012		14.559	8.101
60	2,389,900		59,037	22,432		24.703	9.386
65	1,129,400		37,873	20,693		33.534	18.322
70+	1,907,800		139,108	33,787		72.915	17.710
TOTAL	109,077,000	3,239,588	1,112,584	1,090,770	29.700	10.200	10.000

RURAL REGION

Age	Population	Births	Deaths	Outmig.	Observed rates (per 1000) Birth	Death	Outmig.
0	64,966,800		3,749,333	360,672		57.712	5.552
5	68,071,500		404,496	269,265		5.942	3.956
10	54,639,700		142,663	193,276		2.611	3.537
15	36,502,000	1,811,182	101,879	415,552	49.619	2.791	11.384
20	32,627,500	4,331,904	138,066	693,277	132.768	4.232	21.248
25	31,843,600	4,308,732	131,882	300,528	135.309	4.142	9.438
30	28,551,700	3,271,086	133,672	174,397	114.567	4.682	6.108
35	26,011,900	2,087,353	147,542	119,837	80.246	5.672	4.607

APPENDIX A Continued

Age	Population	Births	Deaths	Outmig.	Observed rates (per 1000) Birth	Death	Outmig.
40	22,648,400	884,424	165,168	88,172	39.050	7.293	3.893
45	18,315,900	329,073	187,991	65,227	17.967	10.264	3.561
50	16,879,800		265,956	43,696		15.756	2.589
55	10,432,000		229,172	49,269		21.968	4.723
60	11,944,300		445,211	61,358		37.274	5.137
65	5,691,800		287,999	56,601		50.599	9.944
70+	9,629,600		1,059,459	92,417		110.021	9.597
TOTAL	438,756,500	17,023,754	7,590,489	2,983,544	38.800	17.300	6.800

ALL OF INDIA

Age	Population	Births	Deaths	Outmig.	Observed rates (per 1000) Birth	Death	Outmig.
0	79,107,000		4,290,163	492,532		54.232	6.226
5	82,869,800		462,774	367,707		5.584	4.437
10	68,277,200		166,261	263,937		2.435	3.866
15	47,466,900	2,172,377	122,124	567,476	45.785	2.574	11.960
20	43,082,400	5,255,111	167,386	946,736	121.978	3.885	21.975
25	40,799,300	5,114,688	156,463	410,400	125.362	3.835	10.059
30	36,164,100	3,851,137	157,292	238,156	106.491	4.349	6.585
35	32,893,400	2,454,628	173,410	163,649	74.624	5.272	4.975
40	28,362,700	1,032,836	192,786	120,407	36.415	6.797	4.245
45	22,792,400	382,565	218,441	89,074	16.785	9.584	3.908
50	20,690,100		305,743	59,671		14.777	2.884
55	12,655,400		261,543	67,281		20.667	5.316
60	14,334,200		504,248	83,790		35.178	5.845
65	6,821,200		325,872	77,294		47.773	11.331
70 +	11,537,400		1,198,567	126,204		103.885	10.939
TOTAL	547,833,500	20,263,342	8,703,073	4,074,314	36.988	15.886	7.437

SOURCE: Rogers (1982b)

APPENDIX B: Multistate Population Flows: Sweden, Females, 1974

To / From	Stockholm				Rest of Sweden						
	Never Married	Married	Widowed	Divorced	Never Married	Married	Widowed	Divorced	Deaths	Births	Population
Stockholm											
Never married	–	7,130	–	–	7,648	352	–	–	1,280	2,827	304,024
Married	–	–	4,408	9,312	–	5,748	50	48	2,123	6,858	363,581
Widowed	–	74	–	–	–	0	395	–	2,966	25	67,847
Divorced	–	1,238	–	–	–	63	–	422	271	281	31,113
Rest of Sweden											
Never married	6,826	1,257	–	–	–	29,029	–	–	6,282	12,430	1,355,406
Married	–	3,744	206	169	–	–	20,297	18,437	9,724	29,661	1,526,855
Widowed	–	0	199	–	–	325	–	–	14,831	94	317,223
Divorced	–	204	–	253	–	4,837	–	–	1,277	1,024	132,486
Sweden total									38,754	53,200	4,098,535

SOURCE: Rogers and Planck (1984).

REFERENCES

Berelson, B. 1974. *World population: status report 1974.* Reports on Population/Family Planning, No. 15. New York: The Population Council.

Coale, A. 1969. Population and economic development. In *The population dilemma,* ed. P. M. Hauser, 2nd ed., pp. 59-84. Englewood Cliffs, NJ: Prentice-Hall.

Coale, A., and Demeny, P. 1966. *Regional model life tables and stable populations.* Princeton University Press.

———. 1983. *Regional model life tables and stable populations.* 2nd ed. New York: Academic.

Frejka, T. 1973. *The future of population growth: alternative paths to equilibrium.* New York: John Wiley.

Keyfitz, N. 1971. Models. *Demography* 8(4): 571-80.

———. 1977. *Applied mathematical demography.* New York: John Wiley.

———. 1980. Do cities grow by natural increase or by migration? *Geographical Analysis* (12(2): 142-56.

Keyfitz, N., and Beekman, J. A. 1984. *Demography through problems.* New York: Springer-Verlag.

Land, K., and Rogers, A., eds. 1982. *Multidimensional mathematical demography.* New York: Academic.

Ledent, J. 1980. *Rural-urban migration, urbanization, and economic development.* Working Paper No. 80-19. Laxenburg, Austria: International Institute for Applied Systems Analysis.

———and Rogers, A. 1979. *Migration and urbanization in the Asian Pacific.* Working Paper No. 79-51. Laxenburg, Austria: International Institute for Applied Systems Analysis.

Littman, G., and Keyfitz, N. 1977. *The next hundred years.* Working Paper No. 101. Cambridge: Center for Population Studies, Harvard University.

Long, J. F. Survey of federally produced national level projections. *Review of Public Data Use* 23: 712-22.

Philipov, D., and Rogers, A. 1981. Multistate population projections. *IIASA Reports* 4(1): 51-82.

Rogers, A. 1968. *Matrix analysis of international population growth and distribution.* Berkeley: University of California Press.

———. 1971. *Matrix methods in urban and regional analysis.* San Francisco: Holden-Day.

———. 1975. *Introduction to multiregional mathematical demography.* New York: John Wiley.

———. 1978a. Model migration schedules: an application using data for the Soviet Union. *Canadian Studies in Population* 5: 85-98.

———. 1978b. Migration, urbanization, resources, and development. In *Alternatives for growth: the engineering and economics of natural resources development,* eds. H. McMains and L. Wilcox, pp. 149-217. Cambridge, MA: Ballinger.

———. 1980. Introduction to multistate mathematical demography. *Environment and Planning A* 12(5): 489-98.

———. 1981. *Projections of population growth and urbanization for five southeast Asian Pacific nations.* Working Paper No. 81-137. Laxenburg, Austria: International Institute for Applied Systems Analysis.

———. 1982a. Sources of urban population growth and urbanization, 1950-2000: a demographic accounting. *Economic Development and Cultural Change* 30(3): 483-506.

———. 1982b. The migration component in subnational population projections. In *National migration surveys, survey mannual X: guidelines for analysis,* pp. 216-254. Bangkok: Economic and Social Commission for Asia and the Pacific, United Nations.

———and Ledent, J. 1976. Increment-decrement life tables: a comment. *Demography* 13(2): 287-90.

Rogers, A., and Philipov, D. 1980. Multiregional methods for subnational population projections. *Sistemi Urbani* 2(2/3): 151-70.

Rogers, A., and Planck, F. 1984. *Parametrized multistate population projections.* Working Paper No. 84-1. Boulder: Population Program, Institute of Behavioral Science, University of Colorado.

Ross, J. A. 1982. Life tables. In *International encyclopedia of population,* ed. J. A. Ross. pp. 420-425. New York: The Free Press.

United Nations. 1976. *Global review of human settlements: a support paper for Habitat.* Oxford: Pergamon.

———. 1980. *Patterns of urban and rural population growth.* Population Studies, No. 68. New York: Department of International Economic and Social Affairs, United Nations.

———. 1981. *World population prospects as assessed in 1980.* Population Studies, No. 78. New York: Department of International Economic and Social Affairs, United Nations.

———. 1984. *Population projections: methodology of the United Nations.* Population Studies No. 83. New York: Department of International Economic and Social Affairs, United Nations.

United States Bureau of Labor Statistics. 1982. *Tables of working life.* Bulletin 2135. Washington, DC: Department of Labor.

Willekens, F. 1979. *Matrix models of aggregate multiregional population change: a comparison.* International Institute for Applied Systems Analysis, Laxenburg, Austria (unpublished paper).

Willekens, F., and Philipov, D. 1981. *Dynamics of multiregional population systems: a mathematical analysis of the growth path.* Working Paper No. 81-75. Laxenburg, Austria: International Institute for Applied Systems Analysis.

World Bank. 1979. *World development report, 1979.* New York: Oxford University Press.

———. 1983. *Short-term population projection 1980-2020 and long-term projection 2000 to stationary state, by age and sex for all countries of the world.* Washington, DC: Policy and Research Unit, Population, Health and Nutrition Department, World Bank.

ABOUT THE AUTHOR

ANDREI ROGERS is Professor of Geography and Director of the Population Program, Institute of Behavioral Science at the University of Colorado, Boulder. Since obtaining his Ph.D. in city and regional planning at the University of North Carolina, Chapel Hill, Rogers has also held faculty appointments in the Department of City and Regional Planning at the University of California, Berkeley, and the Department of Civil Engineering at Northwestern University, Evanston, Illinois. Before his move to Colorado, he spent eight years at the International Institute for Applied Systems Analysis in Laxenburg, Austria, where he headed a research program that addressed global human settlement issues and problems. He is the author of a number of books on population analysis and is currently a member of the editorial boards of several journals in the fields of population, planning, regional science, and mathematical biology. His current teaching and research interests revolve around the quantitative analysis of global patterns of migration, urbanization, and economic development.